The Enigmatic Electron

Fundamental Theories of Physics

An International Book Series on The Fundamental Theories of Physics: Their Clarification, Development and Application

Volume 49

The Enigmatic Electron

by

Malcolm H. Mac Gregor
Lawrence Livermore National Laboratory,
Livermore, California,
U.S.A.

KLUWER ACADEMIC PUBLISHERS
DORDRECHT / BOSTON / LONDON

ISBN 0-7923-1982-6

Published by Kluwer Academic Publishers,
P.O. Box 17, 3300 AA Dordrecht, The Netherlands.

Kluwer Academic Publishers incorporates
the publishing programmes of
D. Reidel, Martinus Nijhoff, Dr W. Junk and MTP Press.

Sold and distributed in the U.S.A. and Canada
by Kluwer Academic Publishers,
101 Philip Drive, Norwell, MA 02061, U.S.A.

In all other countries, sold and distributed
by Kluwer Academic Publishers Group,
P.O. Box 322, 3300 AH Dordrecht, The Netherlands.

Printed on acid-free paper

Printed in the Netherlands

To ELEANOR

who has supported and encouraged

this endeavor for many, many years

Table of Contents

Part II. The Natural Size of the Electron

Part III. The Spectroscopic Electron

Part IV. The Mott Channeling of Finite-Sized Electrons

List of Figures

List of Tables

Preface

The Rationale for the Present Book

Perhaps the most critical problem facing present-day particle physicists is to delineate the relationship between classical and quantum systems. This relationship has many facets. Particle-wave duality is one. The concept of the point particle is another. And the concept of particle mass is yet another. The electron, as the lightest of the charged particles, represents a fundamental "ground state", and many of the essential problems in the murky area between the domains of classical and quantum physics can be brought into focus by studying just this one particle. Thus the present book is centered on questions that arise in connection with the electron, and in particular with its mass, which has remained an unsolved, and indeed almost unexplored, mystery. Each student of physics, beginner and professional alike, has to fashion for himself a way of thinking about the electron. If, after reading this book, the reader views this topic somewhat differently than before, the efforts of the author will have been amply rewarded.

When physicists were confronted with the properties of the electron, they made a conceptual leap into the unknown: they concluded that the electron does not obey classical laws with respect to mechanics (as connected to the spin of the electron), and also with respect to electrodynamics (as connected to the magnetic moment of the electron). This leap was made by postulating that the electron is a point-like object, which at one stroke removed both the spin and the magnetic moment of the electron beyond the range of our classical understanding, where they have remained ever since. If the electron is truly point-like, then the ordinary laws of angular momentum and of electromagnetism clearly do not apply, and we are left at an impasse, both conceptually and mathematically, with respect to these quantities. The conclusion that the electron is point-like was, of course, not based on an arbitrary decision: it seemed to be compelled by the experimental data. The purpose of the present book is to discuss some new results which suggest a re-examination of these ideas.

Although the electron was the first elementary particle to be experimentally identified, and is probably one of the simplest, it is today still a mystery. We know what it *does*, but not what it *is*. Does it have classical features, or is it strictly a quantum phenomenon?

Or, is it perhaps a mixture of properties that we know from the macroscopic world combined with properties that uniquely pertain to the microworld? *Classical physics* deals, by and large, with concepts that can be visualized. Most of us, physicists and non-physicists alike, retain the hope that, when properly understood, *quantum physics*, and its trademark, the elementary particle, will also be expressible in terms of visual concepts. The present studies represent an attempt to advance that understanding.

The thesis of this book is contained in the following assertions: *(a)* a classical approach to electron structure can be pushed much further than is generally recognized; *(b)* the properties of the electron mass are rather sharply defined by relativity theory and experiment; *(c)* explanations that are essentially classical exist for quantum properties such as the gyromagnetic ratio, the quantized spin, and the anomalous magnetic moment of the electron; *(d)* experiments may already exist that reveal a finite size for the electron. In developing this thesis, we will assemble the various facets of our knowledge that can be brought to bear on these topics. Each facet carries with it some uncertainties, but the mosaic they form is highly suggestive.

Our goal in these studies is to establish a new point of view. Hence, especially in the first part of the book, we will sometimes use a spiral approach, treating a topic more than once, each time rather briefly, and each time featuring a different aspect of the problem.

We conclude this discussion with the following precautionary comment. The reader will quickly discover that much of the mathematics contained in this book is, in terms of present-day sophistication, quite elementary. The reason for this is that many of the unanswered questions we are investigating do not require complex mathematics. But just because an equation is simple is no reason to think that it is trivial or unimportant. In fact, the case is likely to be just the opposite. History has demonstrated that simple relationships constitute the underpinning of science. For example, the Einstein equation $E = mc^2$ is surely one of the simplest mathematical expressions in all of physics. And yet, the results that spring from this equation, which has been recognized for less than a century, have transformed not only the way we view the basic elements of the universe, but also the way we make both electricity and war. It might seem that the elegant and complicated mathematical formalism of *quantum electrodynamics* (*QED*) – the quantum mechanics of the electron as displayed in its interactions with the electromagnetic field – serves as a counterexample to this assertion about the importance of elementary equations. However, although *QED* tells us with astounding accuracy *how* the electron interacts, it unfortunately conveys very little information as to the nature of the electron itself, other than the important fact that its electrical charge is point-like. When we undertake the task of reconstructing the spectroscopic properties of the electron, we discover that much of the information we can bring to bear on this problem is contained, as we will discuss in detail, in rather pedestrian mathematics.

Most of the equations used in this book are familiar to physicists, but the conclusions we draw from discussions of these equations are in some cases unfamiliar. Moreover, there are some equations derived here, in particular those of the relativistically spinning sphere, which are not to be found in any of the textbooks of physics.

Inspiration for these studies sprang in no small measure from the introduction to physics that I and my fellow postwar students received from Professor George Uhlenbeck at the University of Michigan in the late 1940s and early 1950s. He indoctrinated us in the mysteries of electricity and magnetism and also of quantum mechanics. And, as one of the principal architects of the electron, he was our model of a theoretical physicist. He also gave me the honor of serving as a member of my Ph.D. thesis committee. I would like to think that the ideas in this book represent in some sense an affirmation of the concept of the spinning electron that Professor Uhlenbeck and his colleague, Dr. Samuel Goudsmit, presented to the world. In the development of these ideas, the existence of journals such as *Il Nuovo Cimento*, *Lettere al Nuovo Cimento* and *Foundations of Physics Letters*, in which unorthodox research can find a haven, has been indispensable. Thanks are due to Stephen Warshaw and Richard White for useful comments on the manuscript. Finally, I should note that this book could not have been written without the unfailing help I have received from my wife and children, who have shared with me the vicissitudes of more than two decades of work in trying to unravel the enigma of the electron.

Physics Department
University of California
Lawrence Livermore National Laboratory
Livermore, California

MALCOLM MAC GREGOR
July 1992

Part I

The Crisis: Classical *Vis à Vis* Quantum Physics

"You know, it would be sufficient to really understand the electron."

Albert Einstein[1]

Chapter 1

Three Unanswered Questions in Twentieth Century Physics

The science of physics, after progressing slowly for a couple of millenia, suddenly exploded in the eighteenth and nineteenth centuries with the development of the classical domain, and exploded again in the twentieth century with the development of the quantum domain. The results of these endeavors have transformed civilization. Hence, we might logically suppose that the main edifice of physics has now been erected, and only some fleshing out of the details remains to be accomplished. However, some very basic – one might even say crucial – questions remain unanswered, questions that threaten to modify some of the underpinnings of this edifice. In the present book we examine three of these questions in detail. And, as a means of sharpening this examination, we couch these questions in a very specific manner:

I. What is the size of the electron?

II. What is the nature of the electron mass?

III. Are there experiments that reveal the size of the electron?

These questions are of importance in that they influence not only the way we calculate, but also the way we think. Although these questions are centered on the electron, the information we obtain while trying to answer them also applies to other particles, and hence has a much broader significance.

The questions we have just set forth are of course questions that have been considered at one time or another by all students of physics. Thus, we must explain why we are raising them again, and, in particular, what we have to offer the reader in the way of new information. In this chapter, we briefly sketch the reasons why these questions, in the opinion of the author, are still unanswered. Then, in Chapter 2, we mention some new results that can be brought to bear on these topics. The detailed exposition of these ideas constitutes the remainder of the book.

In the broadest sense, the relevance of an examination of the properties of the electron stems from its bearing on the relationship between classical and quantum physics. As pointed out for example by Zurek (1991), present-day scientists tend to divide the world of physics into two domains, each having its own "theory of physics": the *macroscopic* domain, in which classical concepts prevail; and the *microscopic* domain, in which the precepts of quantum mechanics rule. Zurek describes this situation as follows:

"In the absence of a crisp criterion to distinguish between quantum and classical, an identification of the 'classical' with the 'macroscopic' has often been tentatively accepted."

The electron itself clearly belongs to the microscopic domain. It thus provides us with an opportunity to see if classical concepts still make sense when we attempt to apply them in this domain. This is the thrust of the present studies, and it represents a relatively unexplored field of endeavor in recent times. But there has recently been real progress in the other direction; that is, in applying quantum concepts to the macroscopic domain. Again quoting Zurek:

"The inadequacy of this approach (*classical* = *macroscopic* and *quantum* = *microscopic*) has become apparent as a result of relatively recent developments. A cryogenic version of the Weber bar – a gravity-wave detector – must be treated as a quantum harmonic oscillator even though it can weigh a ton."

As Zurek argues in his article, the line of demarcation between classical and quantum systems cannot be drawn solely on the basis of the sizes of the objects involved – macroscopic or microscopic. Rather, it has to do with decoherence factors that stem from the complete system – the *physical experiment* plus the *surrounding environment*. The decoherence effects profoundly affect the wave function of the system, and hence its quantum nature.

The topic of decoherence is also discussed by Omnés (1992), in the context of a comprehensive review article on the interpretation of quantum mechanics. As Omnés points out, decoherence arises from the dissipative effects that operate on a system as a consequence of its interaction with an environment which has too many degrees of freedom. This dissipation eliminates the off-diagonal wave-function correlations between the various possible states of the system, so that the final state emerges as a classical probability distribution. The decoherence times for typical systems can be estimated, and they are usually very short.[2] The Weber gravity-wave bar mentioned above, which has been carefully decoupled from its environment, has very small dissipation, and hence a long decoherence time.

The important conclusion for our purposes is that if the *size* criterion for separating "classical" from "quantum" systems has now been invalidated – that is, if this line of demarcation between the classical and quantum domains has been breached – then the idea of attempting to extend classical concepts down to microscopic sizes has acquired a new sense of legitimacy. The usefulness of this attempt depends, of course, on the results that are thereby obtained. Ultimately we must arrive at a single physical theory which embraces the viable features of both classical and quantum physics.

We now consider in turn the three questions raised above, and we also briefly discuss an Electron Workshop that was held in 1990 in Antigonish, Nova Scotia. This workshop is the first scientific gathering devoted solely to the electron since its discovery almost a century ago.

A. What is the Size of the Electron?

The electron was the first elementary particle to be discovered, and its properties have been exhaustively investigated. Hence the question as to its size is one that seemingly should have been decided long ago. And, indeed, in the minds of most present-day physicists, this question has already been decided: *the electron is a point-like particle* – that is, a particle with no measurable dimensions, at least within the limitations of present-day instrumentation. However, a rather compelling case can be made for an opposing viewpoint: namely, that *the electron is in fact a large particle which contains an embedded point-like charge*. It is of considerable interest to marshal the evidence that points to this latter view, which is a task we will undertake.

In raising the question as to the size of the electron, we do not have in mind relatively small factors such as $\sqrt{3}$. Rather, we are talking here of length scales that differ by six orders of magnitude! (We do not include the extremely small sizes at which gravitational effects become important.) We make this remark quantitative by defining seven different "sizes" that can be ascribed to the electron. These are characterized by the following radii:

$$R_C = \text{Compton radius}, \tag{1.1a}$$

$$R_{QMC} = \text{quantum-mechanical Compton radius}, \tag{1.1b}$$

$$R^{\alpha}_{QMC} = \text{QED-corrected quantum-mechanical Compton radius}, \tag{1.1c}$$

$$R_E = \text{radius of the electric charge on the electron}, \tag{1.1d}$$

$$R_{QED} = \text{observed QED charge distribution for a bound electron}, \tag{1.1e}$$

$$R_0 = \text{classical electron radius}, \tag{1.1f}$$

$$R_H = \text{magnetic field radius}. \tag{1.1g}$$

We now briefly describe each of these radii in turn.

If we view the electron classically, then its mass and spin angular momentum combine together, as is described in Chapter 6, to mandate that its radius must be comparable to the electron Compton radius,

$$R_C = \frac{\hbar}{mc} = 3.86 \times 10^{-11}\,\text{cm}, \tag{1.2}$$

where $\hbar = h/2\pi$, and m is the observed (spinning) mass of the electron in its rest frame. Furthermore, as we demonstrate in Chapter 13, the formalism of quantum mechanical spin and magnetic moment projection factors leads to an electron radius that is a factor of $\sqrt{3}$ larger than R_C, so that

$$R_{QMC} = 6.69 \times 10^{-11}\,\text{cm}. \tag{1.3}$$

And, if we apply magnetic self-energy corrections, as discussed in Chapter 8, we must further increase R_{QMC} by a factor of $(1 + \alpha/2\pi)$, so that

$$R^{\alpha}_{QMC} = 6.70 \times 10^{-11}\,\text{cm}. \tag{1.4}$$

However, the scattering properties of the electron mandate a vastly smaller radius for its electric charge (Bender, 1984),

$$R_E < 10^{-16}\,\text{cm}\,, \tag{1.5}$$

which is commonly accepted today as the radius that characterizes the *actual size of the electron.* With respect to the way we regard the electron, the factor of a million disparity between the radii R_{QMC} and R_E is crucial. If the electron has a radius that is comparable to R_{QMC}, then we can quantitatively reproduce its basic properties in a classical context, which demonstrates that classical physics still applies in this domain. But if it has a radius that is comparable to or smaller than R_E, then none of its basic properties are classically calculable, and the electron itself is removed beyond the reach of our classical comprehension. This latter viewpoint is almost universally regarded as correct today, and it leaves us with no visual or calculational explanation for either the magnetic moment or the spin value of the electron, nor for the gyromagnetic ratio g of these two quantities.[3] In particular, it has rather recently been recognized that not even the Dirac equation can be invoked to resolve this conceptual dead end. Gottfried and Weisskopf, in their book *Concepts of Particle Physics, Volume 1*, comment on this situation as follows:[4]

"At one time it (the Dirac equation) was thought to 'explain' the spin s = 1/2 of the electron, but we now know that this is not so. Equations of the Dirac type can be constructed for any s. At this time we have no understanding of the remarkable fact that the fundamental fermions of particle physics (electrons, neutrinos, quarks, etc.) all have spin 1/2."

Although the electric charge e of the electron, as viewed in scattering experiments, seems to be point-like, its manifestation in atomic bound states is not point-like. The Lamb shift reveals that the electric charge is smeared out over a region of space which is comparable to the electron Compton radius R_C, as we discuss in Chapter 7. This smearing out is attributed to two characteristic phenomena of quantum electrodynamics: *vacuum polarization*, which broadens the electric field of the charge; and *zitterbewegung*, which broadens the spatial location of the charge. Thus we have an effective electron bound-state QED charge radius

$$R_{QED} \cong R_C\,. \tag{1.6}$$

There are two other sizes that we can logically associate with the electron. If we assert that its mass arises from the classical self-energy of the *electric* charge, we arrive at a characteristic charge radius

$$R_O = \frac{e^2}{mc^2} = 2.82 \times 10^{-13}\,\text{cm}\,, \tag{1.7}$$

which is known as the *classical electron radius*. This is discussed in Chapters 3 and 7. Alternately, if we assert that the electron mass arises, at least in part, from the classical self-energy of its *magnetic* field, then we are led to a magnetic radius

$$R_H > 4 \times 10^{-12}\,\text{cm}\,, \tag{1.8}$$

as is derived in Chapter 8. These various types of electron radii are displayed together in Figure 1.1. One of our assignments in the present studies is to develop an electron model that can account for these disparate values of the electron radius.

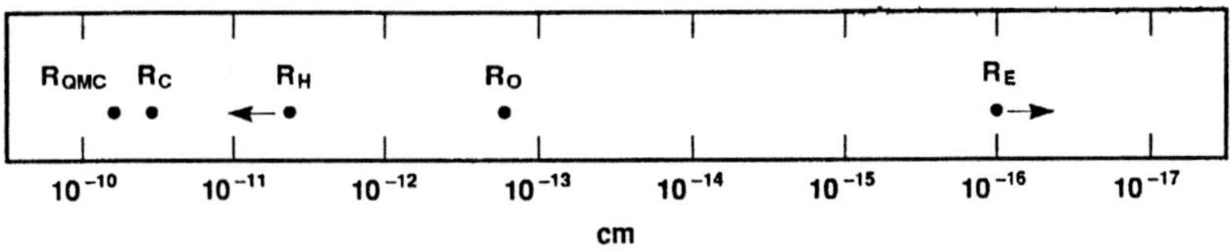

Fig. 1.1. Electron radii, as defined in Eqs. (1.1) - (1.8).

B. What is the Nature of the Electron Mass?

This is the second question listed at the beginning of the chapter, and is a query as to the physical composition of the mass of the electron. It is, somewhat surprisingly, perhaps the most overlooked, or at least bypassed, question in the physics of the twentieth century. When the electron was first discovered, the logical assumption was made that its mass is purely electromagnetic, and the early models of the electron pictured it as a spatially extended electric charge distribution. However, as we discuss in Chapter 3, this viewpoint ran into both experimental and conceptual difficulties. In particular, no combination of purely electric and magnetic charges is stable: an electromagnetic electron should explode. Also, the self-energy that arises from an extended charge distribution, when combined with the small radius that we now know corresponds to electron scattering experiments, leads to an electrostatic self-energy that is much larger than the observed mass of the electron. Thus it was clear within the first two decades after its discovery, and is even clearer now, that the electron is *not* a purely electromagnetic entity. It must have a non-electromagnetic mass component. But what is this non-electromagnetic mass?

In his book *'Subtle is the Lord...'*, which is a book on the life of Albert Einstein, Abraham Pais devotes a section to the topic *Electromagnetic Mass: The First Century.* The first paragraph of this section concludes as follows:[5]

"The electromagnetic mass concept celebrates its first centennial as these lines are written. The investigations of the self-energy problem of the electron by men like Abraham, Lorentz, and Poincaré have long since ceased to be relevant. **All that has remained from those early times is that we still do not understand the problem.**" (Emphasis added.)

The final two sentences of this section reinforce this plaint:[5]

"Recently, unified field theories have taught us that the mass of the electron is certainly not purely electromagnetic in nature.

But we still do not know what causes the electron to weigh." (Emphasis added.)

The electron is now universally recognized as containing a *non-electromagnetic* mass component, which we will henceforth designate as a *mechanical* mass. The electromagnetic aspects of the electron have been analyzed in exhaustive detail, but the mechanical aspects have been essentially ignored. It does not seem to be recognized that if we postulate the existence of a mechanical electron mass (as we must), and if we then use experiment and relativity theory as guides in ascertaining the properties of this mechanical mass (as we should), we are led rather firmly to some surprising and quite remarkable conclusions.

In particular, we are essentially forced into the discovery of the relativistically spinning sphere, which at one stroke ties together the main spectroscopic features of the electron, as we discuss in Chapters 10 and 11. The mechanical mass itself is a strictly microscopic phenomenon that has no counterpart in the macroscopic world. Thus properties such as its rigidity and its interactions with other masses are unknown. It is not clear if we can directly investigate these properties experimentally. However, as we discuss in Chapter 15, we can use phenomenological arguments to guess what these properties must be in order to account for the observable aspects of the electron.

C. Are There Experiments that Reveal the Size of the Electron?

This question, the third one listed at the beginning of the chapter, may have a rather surprising answer. There seem to be such experiments, and indeed the answer may already exist in experimental data that were published decades ago. These experimental effects do not occur in the domain of atomic physics – below a keV of electron energy, nor in the domain of particle physics – above an MeV. They occur (if our assumptions here are correct) in a narrow band of energies in the kilovolt range. The effect in question is a *channeling effect* that can occur (for appropriate impact parameters) in the Mott scattering of electrons and positrons off atomic nuclei. It arises from the spiral trajectory of the equatorial charge on the spinning electron or positron, and it results in an enhanced forward peaking of the elastic scattering. This is essentially a classical effect. As described in Chapter 16, a detailed computer analysis was carried out to delineate the energy window for this channeling process. There are some experimental data on Mott scattering which may indicate the existence of this channeling, as we discuss in Chapter 17. However, the data are not definitive, and new experiments are required in order to clearly resolve this issue.

D. The 1990 Antigonish Electron Workshop

The electron was discovered during the 1890s, as we describe in Chapter 3. In August of 1990, almost a hundred years later, a group of experimental and theoretical physicists gathered together at St. Francis Xavier University in Antigonish, Nova Scotia, to hold an Electron Workshop. The Proceedings of this workshop have been published in another volume of this series (Hestenes and Weingartshofer, 1991).[6] It is remarkable that the Antigonish Workshop is the first conference ever held that was devoted entirely to the electron. And it is even more remarkable that the *raison d'être* for this workshop was to discuss the serious scientific problems posed by the electron. How can this most simple and most studied of particles still elude our scientific grasp? In his lead-off article from the Antigonish Workshop, Jaynes (1991, p. 1) describes this situation, and he gives us the answer:

"We are gathered here to discuss the present fundamental knowledge about electrons and how we might improve it. On the one hand it seems strange that this is the first such meeting, since for a Century electrons have been the most discussed things in physics. And for all this time a growing mass of technology has been based on them, which today dominates every home and office. But on the other hand, this very fact makes it seem strange that a meeting like this could be needed. How could all this marvelously successful technology exist unless we already know all about electrons?

The answer is that technology runs far ahead of real understanding."

In considering the problems posed by the electron, it is important to distinguish between the topics covered in the Antigonish Workshop and the topics treated in the present book. The electron presents physicists with two general kinds of problems: *(a)* the problems associated with the *particle* aspects of the electron; *(b)* the problems associated with the *electron wave.* In the present studies, we are considering only problems of type *(a).* To us, the electron *is* a particle, and we are seeking to learn all we can about it as a localized entity that moves in localized trajectories. In the Antigonish meetings, both the type *(a)* and type *(b)* problems were taken up. This expansion of the agenda does far more than merely increase the number of puzzling difficulties. It opens up a whole new class of difficulties – namely, the problems of particle-wave duality. In the Schrödinger formulation of quantum mechanics, for example, the electron wave contains the entire information about the behavior of the electron, and the electron itself doesn't even appear. Where has it gone? And in electron virtual-double-slit experiments, we have the following sequence of events: *(1)* a single electron is emitted from the cathode of an electron microscope; *(2)* something (the electron and/or the electron wave) travels from the point of emission to a charged metallicized quartz filament, splits into two parts, moves around *both* sides of the filament under the guidance of the electrostatic field, and subsequently interferes with itself in the region past the filament; *(3)* an electron finally appears at a location and with a probability that corresponds to the calculated interference pattern.[7] As Hestenes (1991, p. 31) puts the matter:

"Is the electron a particle always, sometimes, or never? Theorists have come down on every side of this question."

By limiting ourselves in *The Enigmatic Electron* to the particle aspects of the electron, we are bypassing the difficulties inherent in the electron wave. However, if we can develop a comprehensive model for the electron as a localized and visualizable entity, this information may eventually carry over and tell us something about the features of the electron wave.

In an early review article on the quantum theory of radiation, Fermi (1932) made the following summary:

"In conclusion, we may therefore say that practically all the problems in radiation theory which do not involve the structure of the electron have their satisfactory explanation, while the problems connected with the internal properties of the electron are still very far from their solution."

Almost sixty years later, Barut (1991, p. 109) restated these difficulties in more detail:

"If a spinning particle is not quite a point particle, nor a solid three dimensional top, what can it be? What is the structure which can appear under probing with electromagnetic fields as a point charge, yet as far as spin and wave properties are concerned exhibits a size of the order of the Compton wave length?"

This quotation by Barut in fact pinpoints the focus of the present studies. The electron model that we develop in Chapters 9-14 pulls together these disparate properties of the electron, and thus provides an answer to Barut's question. It is beyond the scope of the present discussion to go into detail about the various electron papers delivered at the Antigonish Workshop (Hestenes and Weingartshofer, 1991). What is relevant here is to note that none of the talks presented at this workshop bear directly on the three questions listed at the beginning of this chapter.[6] Although many of the participants at the workshop raised serious doubts about currently held beliefs with respect to the electron and to the quantum mechanics of electron systems, none of them ventured along the lines we follow in the present studies. Theoretically, considerable work was presented at the Antigonish Workshop on the use of the Dirac equation to deduce Compton-sized *zitterbewegung* effects.[8] Experimentally, the implementation of high-intensity laser and microwave electromagnetic field techniques promises to open up new avenues for the detailed exploration of electron systems.[9]

The above discussion characterizes the general tenor of the present book, which is in reality a monograph on the electron. Although the electron is generally considered to be a point-like particle whose properties cannot be understood classically, we will examine the case for a different *ansatz*: namely, that the electron has a large radius, comparable to its Compton radius R_C, and that its basic properties, including its first-order anomalous magnetic moment and its point-like scattering behavior, can be accurately reproduced by what is in many respects a classical model. Since the electron represents one of the most-studied topics in twentieth-century physics, we must, in order to justify the present work, delineate the new ideas that can be added to these studies. This is the subject of the next chapter.

Notes

1 The quotation by A. Einstein shown on page 1 is from Barut (1991, p. 108).

2 Omnés (1992) discusses Schrödinger's cat (Schrödinger, 1935) as an example in which decoherence quickly reduces the problem to one of classical probabilities.

3 The gyromagnetic ratio g is defined as g = (magnetic moment)/(spin)/($e/2mc$), and it has the (approximate) value $g = 2$ for the electron.

4 See Gottfried and Weisskopf (1984, p. 38).

5 See Pais (1982, pp. 155 - 159).

6 The Proceedings of the 1990 Antigonish Electron Workshop (Hestenes and Weingartshofer, 1991) were not available to the author until the present text had already been completed in draft form and submitted for publication.

7 See Mac Gregor (1988).

8 Articles by Jaynes (p. 1), Hestenes (p. 21), Gull (p. 37), Krüger (p. 49), Boudet (p. 83) and Barut (p. 105) in Hestenes and Weingartshofer (1991) treat various aspects of zitterbewegung.

9 An article by Weingartshofer (p. 295) in Hestenes and Weingartshofer (1991) gives an overview of laser-electron interactions.

Chapter 2

Some New Ideas in an Old Field of Physics

In this chapter we briefly summarize some new ideas that have relevance to the topic of *The Enigmatic Electron.* The electron represents an old and exhaustively examined field of physics, and without new ideas there would be little point in going over it again. As an inducement to the reader, we gather here in one place the various innovations that we feel can and should be added to the accepted lore on this topic. There are a surprising number of these, in spite of the voluminous literature that already exists on the subject. The main reason for this circumstance is that most physicists long ago gave up on attempts to seek a classical understanding of the electron. Rohrlich (1965, p. vii) describes the situation as follows:

"With very few notable exceptions the classical theory of charged particles has been largely ignored and has been left in an incomplete state since the discovery of quantum mechanics. Despite the great efforts of such men as Lorentz, Abraham, Poincaré, and more recently Dirac, it is usually regarded as a "lost cause". This is deplorable indeed. I feel that with the present state of our knowledge we are able to look back at this theory, to put it in its right perspective, and, above all, to *complete* the unfinished work of the past."

And Pearle (1982, p. 213) addresses this same topic:

"The state of the classical electron theory reminds one of a house under construction that was abandoned by its workmen upon receiving news of an approaching plague. The plague in this case, of course, was quantum theory. As a result, classical electron theory stands with many interesting unsolved or partially solved problems. Occasional workmen have since passed by and added a bit, most notably to the theory of the point electron. But one cannot really understand the present state of the theory without some knowledge of its genesis."

In Chapter 3 we will give a brief survey of some aspects of this genesis.

When writing a book in a nonfiction field such as physics, the author usually has one of two purposes in mind: to review and summarize the existing and generally accepted body of knowledge; or to seek to extend this body of knowledge by presenting new material and new ideas, including possible reinterpretations of old ideas. In the present discussion, our goal is the latter. Many excellent texts and monographs have been written on various aspects of the electron.[1] We select from these only the information that is relevant for the present purposes. Thus we do not offer a pedagogical survey of this topic. Rather, we deal with what appear to be some "open questions", and, as far as possible, with material that is limited to ideas which directly relate to these questions.

Many of the ideas that we discuss in this book have been previously published by the author in the refereed literature, but they are not generally known to most physicists. Some of the ideas are presented here for the first time. It is the total ensemble of these intrinsically related ideas that makes a rather compelling case for a re-evaluation of our notions about the electron. This of course is the motivation for the present monograph.

With respect to the *spectroscopic* aspects of the electron, we will develop and discuss the following topics:

(1) A more-or-less classical model exists that correlates the spectroscopic properties of the electron accurately to first order in $\alpha = e^2/\hbar c$.

(2) This model is built on the concept of a relativistically spinning "mechanical" (i.e., non-electromagnetic) sphere of matter[2] *that contains an equatorial point charge e.*

(3) The mechanical mass has two unusual experimentally mandated properties:
(a) it behaves both relativistically and in scattering experiments as a "rigid body";
(b) it is essentially non-interacting (neutrino-like).

(4) The relativistically spinning sphere transforms correctly under Lorentz transformations.[3]

(5) It also accounts in a natural way for the quantization of the electron spin.

*(6) The model reproduces the **observed** electron spin* $J_Z = 1/2\,\hbar$*, and also the **total** quantum-mechanical spin* $J = \sqrt{3}/2\,\hbar$.

*(7) The model reproduces the **observed** electron magnetic moment* $\mu_Z = e\hbar/2mc$*, and also the **total** quantum-mechanical magnetic moment* $\mu = \sqrt{3}e\hbar/2mc$.

(8) It follows from (6) and (7) that this classical model correctly reproduces the gyromagnetic ratio of the electron, and thus stands as a counter-example to the frequently expressed opinion that no such model exists.[4]

(9) The quantum-mechanical spin projection factor $\sqrt{J_Z/(J_Z+1)}$ *<=>* J_Z *causes the electric quadrupole moment of the electron to vanish on the average, so that it is not an observable, but it still plays a key role via the wave function symmetry requirements, and thereby leads to the requisite two-component representation of the rotation group.*

(10) The electrostatic self-energy of the electron in this model is necessarily equal to zero.

(11) The magnetic self-energy[5] *of the electron gives rise to its anomalous magnetic moment,*

$$\mu_A \equiv \mu\left(1 + \frac{\alpha}{2\pi}\right) ,$$

and the characteristic logarithmic energy divergences of QED appear to correspond mathematically to magnetic rather than electrostatic effects.

These are the spectroscopic features of the model. There are also two crucial *dynamical* (interactive) results that we treat in detail in Part IV:

(12) *Although this electron model is large, with a radius that is comparable to the electron Compton radius $R_C = \hbar/mc$ (or, more precisely, $R_{QMC} = \sqrt{3}\hbar/mc$), it can reproduce the fact that, at most energies, electron scattering occurs in a point-like manner.*

(13) *There is an energy window in the keV region in which the electron may not be exhibiting point-like scattering,[6] thus providing another possible counter-example to prevailing belief.*

The spherical *mechanical* electron mass that we introduce in these studies has some interesting and unusual features:

(a) *In the electron, the mechanical mass is spinning at the relativistic limit, with its equator moving at, or infinitesimally below, the velocity of light, c.*

(b) *The calculated relativistic mass increase, from either special or general relativity, is finite, and is equal to a factor of 3/2: $m_S = 3/2\ m_0$.*

(c) *The calculated relativistic moment-of-inertia is $I = 1/2\ m_S R^2$.*

(d) *Since both the angular distributions and the absolute values of $e^- e^-$ (Møller) and $e^+ e^-$ (Bhabha) scattering are accurately reproduced by QED, the interactions in these scattering processes are purely electromagnetic. Hence the interactions of the **mechanical masses** in these processes must be orders of magnitude smaller than the interactions of the **electric charges**, as mentioned in (3) above.*

(e) *The assumption of rigidity for the electron mass is mandated for two reasons:[7]*

(e1) *A rigid body has no internal strains, which would lead to the wrong relativistic moment of inertia for the spinning sphere, and would in general raise problems with respect to covariance;*

(e2) *In a rigid body, the translational effect of any applied external force is experienced as a force which operates on the center-of-mass of the body. This is the **Golden Rule of Rigid-Body Scattering** described in Chapter 15, and it leads to the result that a large electron will, at most energies, scatter in a point-like manner.*

When the spectroscopic features and the Lorentz transformation properties of this relativistically spinning sphere model are evaluated numerically on a computer, as we describe in Chapter 12, two important conclusions emerge:

(14) *The correct **gyromagnetic ratio** for the electron is obtained only when the sphere is spinning at the relativistic limit.*

(15) *The correct **Lorentz transformations** for the electron are obtained only when the sphere is spinning at the relativistic limit.*

Thus *spin quantization* is a necessary feature of this model.

In the remainder of Part I we examine the conflicts between classical and quantum physics. These are two aspects of what must ultimately appear as a single unified theory. In Part II we discuss what is known about the sizes of not only the electron, but also the other elementary particles. In Part III we develop the spectroscopic features of the electron.

Finally, in Part IV we consider some dynamical aspects of the electron, as revealed in electron-nucleus (Mott) scattering. The present work may in fact raise more questions than it answers, but it will hopefully lead to some new avenues for exploration.

Notes

1 See Lorentz (1952), Heitler (1954), Jackson (1962), Kramers (1962), Rohrlich (1965), Jauch and Rohrlich (1976), Pais (1982), Pearle (1982), Feynman (1985) and Hestenes and Weingartshofer (1991).

2 See Mac Gregor (1970).

3 See Mac Gregor (1985a).

4 In Feynman (1964, Vol. II, p. 34-3), the statement is made that "there is no classical explanation" for the electron g-value of 2. However, the relativistically spinning sphere model of the electron that is described in Chapter 11 of the present book stands as a counter-example to this assertion. It should be noted that this sphere model was discovered (Mac Gregor, 1970) subsequent to the publication of *The Feynman Lectures in Physics.*

5 See Mac Gregor (1989).

6 See Mac Gregor (1992).

7 The concept of a "rigid body" is useful in presenting the ideas of classical mechanics (Synge and Griffith, 1942, Ch. X - XII; Goldstein, 1950, Ch. 4 and 5; Rosser, 1967, pp. 179-181), but it runs into recognized difficulties when special relativistic considerations become important (Goldberg, 1984, pp. 191-196; French, 1968, pp. 27 and 117-119). It is beyond the scope of the present discussion to attempt to deal with those difficulties, except to point out that *(a)* any difficulties raised with respect to causality are limited to the microworld; and *(b)* the "mechanical matter" that we invoke for the mass of the electron is a "continuum" (see Chapter 15) that has no counterpart in the macroscopic world, and hence in our experimental experience.

Chapter 3

The Breakdown of Classical Physics in the Electron

In the preceding chapters we briefly introduced the dichotomy of classical and quantum viewpoints as applied to the electron. If we attempt to construct a classical electron model – that is, an electron model whose spectroscopic properties are calculated in accordance with the classical rules of mechanics and electromagnetism – we are led to a rather large radius for the electron. The radii R_O, R_H, and R_C shown in Figure 1.1 are examples of this. On the other hand, the dynamics of electron-electron scattering and the formalism of quantum electrodynamics seem to mandate the much smaller radius R_E of Figure 1.1. Faced with this difficulty, we are forced to move beyond the confines of classical nineteenth century physics. Moreover, there are other difficulties. The first models of the electron were purely electromagnetic, and they led to problems with respect to stability, and also with respect to special relativity, which appeared at about this same time. Thus classical physics had to be extended, or possibly even renounced, at the level of the electron. This problem, which was clearly defined three quarters of a century ago, has never been completely resolved. *Do we simply do away with classical concepts on the level of the elementary particle, or do we add to them?*

In this chapter we give a short historical survey of the development of present-day ideas with respect to electrons, so that we can pinpoint as clearly as possible the circumstances which led to the apparent termination of classical physics and the plunge into the abyss of the quantum domain. We first consider the electromagnetic features of the electron, and then we discuss the aspects that relate directly to special relativity.

A. Conflicts with Classical Electromagnetism

The electron, the first elementary particle to be discovered, was identified by J. J. Thomson in 1897. Passing a beam of electrons through crossed electric and magnetic fields, he determined their charge-to-mass ratio e/m, which turned out to be a factor of a thousand or more larger than the e/m ratio of the hydrogen atom. In that same year, J. S. Townsend carried out a forerunner of the Millikan oil-drop experiment and demonstrated that the charge on the electron is the same unit charge e that occurs in atomic physics. Thus it became clear that the large e/m ratio for the electron arises from a small mass m, and not from a large charge e. Abraham and Lorentz then proposed models in which the mass of the electron is attributed to the electrostatic self-energy W_E of the charge e, which was assumed to be spherically spread out over a distance characterized by a radius R_O. The self-energy of such a charge distribution is (Rohrlich, 1965, p. 125)

$$W_E = \mathrm{A} e^2 / R_O, \tag{3.1}$$

where A is a numerical constant of order unity.[1] Setting $W_E = mc^2$ and A = 1, we obtain

$$R_O = e^2/mc^2 = 2.82 \times 10^{-13}\,\mathrm{cm}, \tag{3.2}$$

which is denoted as the *classical electron radius*. Is this a reasonable radius for the electron? It seemed so at the time (circa 1900), especially after it was demonstrated that the scattering of light by an electron is characterized by the Thomson cross section (Jackson, 1962, pp. 489-491),

$$\sigma_T = \frac{8}{3}\,\pi R_O^2. \tag{3.3}$$

However, although this might at first glance appear to be a success for classical physics, it in fact raised a serious problem with respect to the containment of the electric charge. Like-sign electric charge elements repel one another, so that an additional attractive force is required in order to hold the electron together. Furthermore, no combination of purely electric forces, or electric and magnetic forces (as introduced in the next paragraph), can lead to a stable configuration (Oppenheimer, 1970, pp. 85-89). Thus a non-electromagnetic so-called *Poincaré force* (Jackson, 1962, pp. 592-593) is required in order to keep the electron from blowing up. Hence the electric properties of the electron, as viewed in the context of classical electromagnetism, led to a serious conceptual inconsistency. These electric properties, which are essentially *static* in nature, indicate that the structure of the electron in some manner transcends the bounds of classical electromagnetic theory. The electron is *not* a purely electromagnetic object. It has what we can denote as a "Poincaré mass component", whose properties are not delineated by conventional classical theories. The concept of the Poincaré mass is discussed in the next section of this chapter, where it is examined within the context of special relativity.

New difficulties with respect to the nature of the electron arose when its *dynamical* properties were taken into account. The Bohr model of the atom had succeeded in accounting for a great many properties of atomic structure. But one phenomenon that it could not explain was the pervasive doubling of many of the atomic spectral lines in an external magnetic field. The key to the solution of this problem was obtained in 1925 by Uhlenbeck and Goudsmit (Uhlenbeck, 1926), who postulated that the electron has an intrinsic spin angular momentum J and an associated magnetic moment μ. The doubling of the spectral lines in a magnetic field suggested that the electron has two possible spin and magnetic moment orientations in the field, with slightly different energies. Hence the observed spin value must be

$$J = \frac{1}{2}\hbar , \tag{3.4}$$

since, according to quantum mechanics, a spin 1/2 particle has two quantized spin orientations. Furthermore, the gyromagnetic ratio g of the spin to the magnetic moment, which is defined as

$$g = (\mu/J) / (e/2mc) , \tag{3.5}$$

has the value $g = 1$ for atomic orbitals (corresponding to the normal Zeeman effect), and it must therefore have a value $g \neq 1$ for the electron (corresponding to an "anomalous" Zeeman effect) in order to produce an observable doublet splitting. Uhlenbeck and Goudsmit chose the value $g = 2$, which, when combined with Eq. (3.4), gives the electron magnetic moment as

$$\mu_O = \frac{e\hbar}{2mc} . \tag{3.6}$$

The Uhlenbeck-Goudsmit model gave the right number of levels, but the calculated splitting was off by a factor of two. The solution to this difficulty was obtained by Thomas, who showed that when Lorentz transformations are properly carried out in a rotating coordinate system, the missing factor of two is obtained (Jackson, 1962, pp. 364-365). This resolved the spectroscopic problems. However, as mentioned above, the introduction of the electron spin and magnetic moment raised additional difficulties with respect to classical models of the electron. Rasetti and Fermi immediately pointed out (Rasetti, 1926) that the existence of a magnetic moment for the electron implies a magnetic self-energy W_H, whose magnitude can be estimated classically in terms of a magnetic radius R_H that is ascribed to the electron. The classical lower bound for R_H turns out to be $R_H \geq 4 \times 10^{-12}$ cm, as we describe in Chapter 8, and it is an order-of-magnitude *larger* than the classical electron radius R_O of Eq. (3.2). Furthermore, if we invoke *Ampere's hypothesis* (Feynman, 1961a, p. 84), and thereby attribute the magnetic moment of the electron to a rotating current, then classical calculations reveal that this rotating current must correspond to an *even larger* Compton-sized electron, whose radius $R_C = 4 \times 10^{-11}$ cm is an order-of-magnitude larger than R_H. Also, the mass and the spin angular momentum of the electron combine together to classically mandate this same Compton-like size for the electron. These results are discussed in Chapter 6. Thus the *dynamical* aspects of the electron, its spin and magnetic moment, pose an even greater strain on our classical viewpoint than do its *static* aspects, its mass and charge. The reason this strain is so acute, of course, is that modern electron scattering experiments have revealed point-like scattering behavior, with a point-like interaction radius R_E that has now been pushed down below 10^{-16} cm (Bender, 1984). The very small experimental electron interaction radius R_E does not seem to be classically reconcilable with the much larger theoretical electron radii R_O, R_H, and R_C, as can be seen graphically in Figure 1.1. Hence the decision was made very early on that the electron is not a classical entity, and physicists have ceased to worry about the problems associated with trying to microscopically reproduce the spectroscopic properties of the electron.

The *static* properties of the electron were discovered at the turn of the century, and its *dynamical* properties were revealed a quarter of a century later. Still another quarter of a century later, in the immediate post-World War II era, two additional and rather subtle QED properties of the electron emerged, both of which correspond to non-classical phenomena. These two new properties of the electron are *(1)* its anomalous magnetic moment, and *(2)* the electrostatic behavior that is exhibited in the Lamb shift. As described above, the electron magnetic moment shown in Eq. (3.6) was initially introduced on an *ad hoc* basis by Uhlenbeck and Goudsmit, and the relativistic Thomas precession made the phenomenology consistent. The subsequent appearance of the Dirac equation provided a theoretical rationale for this work (but see the comment by Gottfried and Weisskopf in Chapter 1), so that both the experimental and the theoretical status of atomic spectroscopy in the late 1920s seemed to be in proper order. However, hyperfine structure measurements made two decades later by Rabi and coworkers (Nafe, 1947) revealed a slight experimental discrepancy with Eq. (3.6), which Breit (1947) explained as arising from a slightly "anomalous" value for the electron magnetic moment. Instead of Eq. (3.6), we now have

$$\mu = \mu_O \left(1 + \frac{\alpha}{2\pi} + \ldots \right), \tag{3.7}$$

as we discuss in more detail in Chapter 8. The anomalous term $\alpha/2\pi$ was first calculated in QED by Schwinger (1948), who with Feynman, Tomonaga, Dyson and others established the set of calculational rules that constitute the theory of quantum electrodynamics. With respect to the present discussion, the significance of QED is that it is a *non-classical* formalism. Hence it is clear from the anomaly in the electron magnetic moment that the electron has important non-classical components. This result follows independently of the static and dynamical properties of the electron mentioned above.

The non-classical QED calculation of the anomalous magnetic moment of the electron was reinforced by a similar QED calculation of the Lamb shift in atomic nuclei. There are two basic ways in which physicists have attempted to determine the charge radius of the electron, and these two ways yield very different answers. The first way is via electron-electron and electron-positron scattering measurements. These lead to the very small charge radius R_E shown in Figure 1.1. The second way is by studying the Lamb shifts of the electron orbitals in atomic nuclei. Non-zero-angular-momentum P-state and higher orbitals have vanishing wave functions at the position of the nucleus, so the electrons in these orbits have no contact with the nucleus, and they appear in Coulomb interactions as point-like particles. This is a consequence of Gauss's Law. However, zero-angular-momentum S-state electrons have non-vanishing wave functions at the position of the nucleus. Thus S-state electrons (to the extent that we can think of them as having defined trajectories) pass repeatedly back and forth through the position of the atomic nucleus. And the result that is experimentally observed is that the S-state energy levels are shifted slightly upwards in energy as compared to the matching P-state energy levels (which, according to the Dirac equation, should have exactly the same energies). These upward-energy S-state *Lamb* shifts denote a slightly weaker attractive Coulomb potential for the S states, and they indicate that the charge on the electron is, at least on the average, spread out over a distance which is large as compared to the size of the atomic nucleus. In fact, from the magnitude of the observed shift, we can deduce that the rms charge distribution of an S-state electron is comparable to the Compton radius of the electron. The conventional explanation for this result (Milonni, 1980, p. 10) is that the electron is indeed a point, but that the radiation reactions induced by (a) the Coulomb field of the atom, or (b) the fluctuations of the vacuum field, or (c) both, cause the electron to be buffeted around, so that its actual average position is spread out over a region of space which is vastly larger than the intrinsic "size" of the electron itself (but which is still much smaller than the dimensions of the atom). This is the phenomenon that in quantum electrodynamics is denoted as *zitterbewegung* (Corben, 1968, p. 97). The significance of the Lamb shift with respect to the present discussion is that the observed Lamb shifts are in accurate agreement with the calculations of QED (Lautrup, 1972), which are based on Feynman diagrams that correspond to *non-classical* electromagnetic phenomena. Thus, although the electron may in fact have some classical properties, it also has some patently non-classical features, including a point-like charge that appears to be spread out over a region of space which is roughly a factor of 10^{18} larger in volume than the volume of the charge itself.

Given the astounding successes of quantum electrodynamics and the point-like behavior of the electron in its scattering interactions, it is easy to understand why classical approaches to electron structure have fallen into disfavor. However, it is the duty of physicists to attempt to bring *all* of the properties of a system, or a particle, or a theory, into one unified picture, and this attempt should not be abandoned as long as there is any prospect of its success. In the case of QED, for example, we know that it is manifestly incomplete, in spite of its many successes, because it does not include the pion. One of the purposes of the present book is to demonstrate by direct calculations that the prospects for a unified picture of the electron are not nearly as bleak as has been commonly assumed. The calculational tools are available, and many of the pertinent calculations have been patiently waiting in the wings.[2]

B. Conflicts with Special Relativity

The electron was discovered, and its mass and charge roughly delineated, in 1897. In 1902 Abraham proposed an electromagnetic model of the electron, pictured as a small rigid spherical shell of charge. Two years later Lorentz proposed a new model of the electron, which was similar to that of Abraham except that the spherical shell was assumed to be flattened along its direction of motion, in accordance with the FitzGerald-Lorentz contraction. These two models each predicted an increase of mass with velocity, but with slightly different velocity coefficients.[3] And they suffered from two common difficulties: (1) they did not reduce to Newton's equations of motion in the low-velocity limit; (2) they were manifestly unstable, since there was nothing to keep the repulsive Coulomb charge distribution from blowing up. A year later, in 1905, Poincaré showed that both of these difficulties could be overcome by the introduction of a non-electromagnetic force (of unknown origin) that canceled out the electrostatic stresses. In that same year, Einstein came out with his special theory of relativity, which properly reduced to the Newtonian limit, and which exhibited a velocity dependence of the mass similar to that of Lorentz. The calculations of Abraham and Lorentz were based on the use of specific electromagnetic models of the electron, but the Einstein formulation showed that the velocity variation of the mass is a very general result which is valid independently of any structural details of the electron. The first experimental measurements of this velocity dependence were made at Gottingen (Abraham's institution) by Kaufmann, and they seemed to show agreement with the Abraham model. However, as these measurements were improved, re-analyzed, and extended by others, the evidence swung in favor of the Lorentz-Einstein viewpoint. But it took a full decade after Einstein's paper on special relativity before the experiments verified the Lorentz-Einstein form of the velocity dependence.

The history of these early efforts at constructing electron models is described in detail in Miller (1981). The key results with respect to the Abraham and Lorentz electron models are as follows. In the Abraham model, the transverse mass of the electron in the limit of small velocities ($\beta \equiv v/c \ll 1$) is (Miller, 1981, p. 60)

$$m_T \;\cong\; \frac{2}{3}\frac{e^2}{Rc^2}\left(1+\frac{2}{5}\beta^2\right) \;, \tag{3.8}$$

where R is the radius of the electron. In the corresponding Lorentz model, the transverse mass in this same limit is (Miller, 1981, p. 74)

$$m_T \;\cong\; \frac{2}{3}\frac{e^2}{Rc^2}\left(1+\frac{1}{2}\beta^2\right) \;. \tag{3.9}$$

The difference between these two velocity dependences is the term $1/2\ \beta^2$ for the Lorentz electron as against the term $2/5\ \beta^2$ for the Abraham electron, which is a 5 to 4 ratio. We can equate the rest mass of each of these electromagnetic electrons to the electrostatic energy of a spherical shell of electric charge, which gives

$$E_0 \;=\; m_0c^2 \;=\; \frac{e^2}{2R} \;. \tag{3.10}$$

Thus we see that the Abraham and Lorentz electrons, in this formulation, each have electromagnetic energies that reduce to $4/3\ m_0c^2$ in the $\beta = 0$ limit. The spurious factor of 4/3 can be eliminated by a proper handling of the basic electromagnetic equations, as has subsequently been demonstrated by a number of workers (Rohrlich, 1965, p. 17).

Pais (1982, pp. 155-159) discusses the Abraham and Lorentz electron models in a formulation that brings out Poincaré's (1905, 1906) contribution quite clearly. He writes the energy and momentum equations for the Abraham electron in the limit of small velocities v as

$$E_{elm} \;=\; \frac{e^2}{2R}\left(\frac{1}{\beta}\ln\frac{1+\beta}{1-\beta}-1\right) \;\cong\; \frac{3}{4}m_0c^2\left(1+\frac{2}{3}\beta^2\right) \;, \tag{3.11a}$$

$$P_{elm} \;=\; \frac{e^2}{2R\beta c}\left(\frac{1+\beta^2}{2\beta}\ln\frac{1+\beta}{1-\beta}-1\right) \;\cong\; m_0 v \;, \tag{3.11b}$$

where $\beta = v/c$ and $m_0c^2 = 2/3\ e^2/R$. The corresponding Lorentz electron gives

$$E_{elm} = \gamma\frac{e^2}{2R}\left(1+\frac{\beta^2}{3}\right) \cong \frac{3}{4}m_0c^2\left(1+\frac{5}{6}\beta^2\right) , \qquad (3.12a)$$

$$P_{elm} = \gamma\frac{4}{3}\frac{e^2}{2R}\frac{\beta}{c} \cong m_0\mathrm{v} , \qquad (3.12b)$$

where $\gamma = 1/\sqrt{1-\beta^2}$. The Einstein special relativistic equations give

$$E = \gamma m_0c^2 \cong m_0c^2\left(1+\frac{1}{2}\beta^2\right) , \qquad (3.13a)$$

$$P = \gamma m_0\mathrm{v} \cong m_0\mathrm{v} . \qquad (3.13b)$$

In this formulation, the Abraham and Lorentz rest masses are each too low by a factor of 3/4. The Lorentz and Abraham velocity dependences shown in Eqs. (3.12a) and (3.11a) are in the same 5 to 4 ratio as were the velocity dependences shown in Eqs. (3.9) and (3.8). The Einstein equations (3.13) do not contain the troublesome factor of 3/4 in the mass term, and they properly reduce to the Newtonian equations for small values of the velocity v. Also, the Einstein equations are completely general with respect to the nature of the mass m_0, which shows that they depend only on the overall energy and momentum content of the electron.

In order to understand Poincaré's contribution to these ideas, it is useful to recast the Lorentz equations (3.12) in terms of the electromagnetic energy-momentum tensor $T_{\mu\nu}$ (Jackson, 1962, p. 385, Eq. 11.134). We will use a schematic representation, in which the tensor itself stands for a spatial integration,

$$T_{\mu\nu}^{(L,0)} \Leftrightarrow \int T_{\mu\nu}^{(L,0)}\, d^3x^{(L,0)} ,$$

and where the superscripts L and 0 denote the laboratory and center-of-mass coordinate systems, respectively. In this notation, the Lorentz equations become (Jackson, 1962, pp. 584-597).

$$E_{elm}^{(L)} = T_{44}^{(L)} = \gamma\,(T_{44}^{(0)} - \beta^2\, T_{33}^{(0)}) , \qquad (3.14a)$$

$$P_{elm}^{(L)} = \frac{i}{c}T_{34}^{(L)} = \gamma\frac{\mathrm{v}}{c^2}(T_{44}^{(0)} - T_{33}^{(0)}) , \qquad (3.14b)$$

where the electron motion is along the 3-axis. Evaluation of the integrations over $T_{44}^{(0)}$ and $T_{33}^{(0)}$ gives

$$T_{44}^{(0)} \Rightarrow \frac{e^2}{2R} = \frac{3}{4}m_0c^2 , \qquad (3.15a)$$

$$T_{33}^{(0)} \Rightarrow -\frac{1}{3}\frac{e^2}{2R} = -\frac{1}{4}m_0c^2 . \qquad (3.15b)$$

Substituting Eqs. (3.15) into (3.14) gives Eqs. (3.12). Poincaré (1905, 1906) observed that if we had $T_{33}^{(0)} = 0$ in Eqs. (3.14), then these equations would have just the form of the Einstein equations (3.13) (apart from a common factor of 3/4), and hence would reduce properly to the Newtonian limit. But $T_{33}^{(0)}$ is an electromagnetic stress term that causes the electron to blow up. Thus, by eliminating this term, Poincaré could at one stroke stabilize the electron and give it the proper transformation equations. (We use a little hindsight here, since the Poincaré and Einstein papers came out at essentially the same time.) The Poincaré stress tensor $P_{\mu\nu}$ (by definition) contains the terms

$$P_{33}^{(0)} = P_{44}^{(0)} = \frac{1}{4} m_0 c^2 \quad . \tag{3.16}$$

Adding these $P_{\mu\nu}$ terms to the corresponding $T_{\mu\nu}$ terms in Eqs. (3.14) gives the Einstein equations (3.13). The nature of the Poincaré confining mechanism was not delineated, but it is clearly non-electromagnetic in nature.

The Poincaré solution to these problems actually caused a misunderstanding, because it seemed to imply that the lack of covariance in the Lorentz model of the electron was a consequence of the electrostatic stresses that exist in the model. However, a proper formulation of the electromagnetic energy and momentum (Jackson, 1962, pp. 594-597; Rohrlich, 1965, pp. 129-134), which goes beyond the work of Abraham and Lorentz, leads to a purely electromagnetic model that is covariant and still contains electrostatic stresses. Hence we now have a relativistic but unstable electron. This in turn might seem to imply that the addition of non-electromagnetic forces will destroy the covariance, but there are various ways in which these non-electromagnetic forces can be added and still maintain relativistic invariance (Pryce, 1938). Thus the *covariance* and the *stability* of the electron are unrelated properties.

Kaufmann (1906) was the first one to point out in print that the Einstein and Lorentz predictions for the velocity dependence of the electron mass are equivalent (Miller, 1981, p. 329). This soon became known as the Lorentz-Einstein velocity dependence. The Kaufmann experiments have been discussed in detail by Cushing (1981) and by Miller (1981). The Kaufmann data, as he himself analyzed them, marginally favored the Abraham velocity dependence. However, Planck (1907) made a somewhat different analysis, taking into account the rather poor vacuum that probably existed in these experiments, and obtained results that marginally favored the Lorentz-Einstein velocity dependence (Cushing, 1981). The first really definitive data were taken by Neumann (1914), and they clearly ruled in favor of Lorentz and Einstein. However, by that time the theory of special relativity was already firmly established on the basis of other results .

The ideas that we have just recapitulated here were all formed in the first few years of the twentieth century. Since that time there has been a great deal of work on electron models, much of it rather abstract in nature. This work is mentioned briefly in Chapter 9. However, we still don't know all of the answers (see the quotations by Gottfried and Weisskopf and by Pais in Chapter 1). The point we want to bring out in the present chapter is *the manner in which the properties of the electron transcend classical nineteenth century physics*. The stability requirements clearly indicate that the electron is not a pure Maxwellian electromagnetic field. It must also contain a confining *mechanical* mass. But the properties of this mechanical mass, as we have outlined in Chapter 2, are not the same as those of the familiar macroscopic mechanical masses (which, in fact, are mostly empty space). The relationships between the charge and the total energy of the electron, between the magnetic field and the total energy of the electron, and between the mass and the spin angular momentum of the electron, lead to various estimates for the size of the electron, all of which are in conflict with the point-like nature of electron-electron scattering experiments. Furthermore, the observed anomaly in the electron magnetic moment, and the Lamb shift anomaly in the effective electron charge distribution, both indicate that QED, which represents a collection of non-classical effects, plays an important role in determining the higher-order small-scale properties of the electron. And special relativity imposes limits on the kinds of classical models that we can devise for the electron.

There are also other non-classical effects with respect to the electron that we have not considered here. The most significant of these is the whole question of *particle-wave duality*. Is the electron a particle, or a wave, or both, or sometimes one and sometimes the other? This is a topic that will be addressed in another monograph, but which is not considered in the present studies. It remains to this day one of the greatest mysteries in science, and it is manifested in its most enigmatic form in the electron virtual-double-slit experiments.[4,5] There are other quantum puzzles as well. Why is the electron spin quantized (Chapter 10)? What is the origin of the quantum-mechanical projection factors that must be applied to the electron spin and electron magnetic moment in order to get the correct atomic coupling properties (Chapter 13)? And what is the nature of the infinities that plague QED (Chapter 8)?

The electron appears to be one of the simplest of the elementary particles. It has no discernible quark substructure and no decay channels. And yet it appears to represent a complete and utter departure from all of the classical notions we have as to the nature of discrete particles. Margenau (1961, p. 6) has described this situation very succinctly:

"... electrons are neither particles nor waves. They are entities which, because of their inaccessibility to immediate observation, have properties which do not allow themselves to be cast into intuitable or visual forms. According to this ... position, which is most widely held today and is in harmony with the formalism of the quantum mechanical theories ..., **an electron is an abstract thing**, no longer intuitable in terms of the familiar aspects of everyday experience, but determinable through formal procedures such as the assignment of mathematical operators, observables, states, and so forth. In sum, the physicist, while still fond of mechanical models wherever they are available and useful, no longer regards them as the ultimate goal and the quintessence of all scientific description; he recognizes situations where the assignment of a simple model, especially a mechanical one, no longer works and where he feels called upon to proceed directly under the guidance of logical and mathematical considerations and at times **with the renunciation of the visual aspects which classical physics would carry into the problem**." (Emphases added.)

This assertion by Margenau that "an electron is an abstract thing" was published in 1961, and it represents the viewpoint that is espoused by an overwhelming majority of present-day physicists. However, it is worthwhile to ponder a sharply dissenting opinion that was set forth by de Broglie in 1976:

"Theoretical physics has made use, for a long time, of abstract representations. ... They are indeed very useful and even almost essential auxiliaries of reasoning. But one must never forget that the abstract representations have no physical reality. Only the movement of elements localized in space, in the course of time, has physical reality."

In Section III of the present book we will construct a spectroscopically accurate model of the electron that is both localized and concrete.

Classical nineteenth century physics can account for many phenomena, even at the level of the elementary particle, but there are clearly some quantum effects which are beyond its purview, as we have summarized in the present chapter. Hence we seemingly end up with two types of physics – classical and quantum. However, as we discuss in the next chapter, the domains of classical and quantum physics may have more in common than is generally realized.

Notes

1 In Eq. (3.1), A = 1/2 for a pure electrostatic surface charge and 3/5 for a uniform spherical volume charge. If Poincaré forces are added in, these become A = 2/3 and 4/5, respectively.

2 See the comments by Rohrlich and Pearle at the beginning of Chapter 2.

3 Instead of using a mass that increases with velocity, present day relativists tend to reserve the term "mass" for the invariant rest mass of the particle. Thus we should more accurately be speaking in terms of *energies* rather than *masses* in the present discussion.

4 In Feynman (1964, Vol. III, pp. 1-1 and 1-10), the following statements were made as part of a pedagogical discussion of electron virtual-double-slit experiments: "We choose to examine a phenomenon which is impossible, **absolutely** impossible, to explain in any classical way, and which has in it the heart of quantum mechanics. In reality, it contains the **only** mystery. We cannot make the mystery go away by 'explaining' how it works. We will just **tell** you how it works. In telling you how it works we will have told you about the basic peculiarities of all quantum mechanics." ... "One might still like to ask: 'How does it work? What is the machinery behind the law?' No one can 'explain' any more than we have just 'explained'. No one will give you any deeper representation of the situation. We have no ideas about a more basic mechanism from which these results can be deduced."

5 For a computer simulation of an electron virtual-double-slit experiment, see Mac Gregor (1988).

Chapter 4

A Possible Reunion of Classical and Quantum Physics

In Chapter 1 we posed three questions about the electron, and in Chapter 2 we listed some new results that relate to these questions. Then in Chapter 3 we briefly sketched the development of the ideas that led to a break with our classical notions about elementary particles. It is clear that there are areas in the elementary particle microworld where the classical boundaries are transcended. Thus we arrive at two different aspects of physics, which correspond to classical phenomena and to non-classical or quantum phenomena, respectively. These of course must ultimately be treated as two aspects of an embracing overall formalism. Is there a way of distinguishing between the classical and the quantum domains, and can we establish links between these domains? As we discussed in Chapter 1, the boundary between classical and quantum concepts can no longer be drawn on the basis of size – *macroscopic* ⇔ *classical* and *microscopic* ⇔ *quantum* – since some macroscopic systems are now known to be quantum mechanical (Zurek, 1991). In the present chapter, we examine another way of dividing the classical and quantum domains, and we discover that this second boundary has also been breached.

The quantum domain has one distinguishing attribute – Planck's constant h. The discovery of this universal constant heralded the beginning of the quantum era. It also marked the entrance of the outstanding puzzle of quantum physics *vis à vis* classical physics – namely, the concept of particle-wave duality. The constant h (or $\hbar = h/2\pi$) characterizes the *quanta* of quantum mechanics, and thus delineates a level of discreteness that is missing in classical formulations. This constant was first identified by Max Planck in connection with the photon – the discrete quantum of electromagnetic particle-wave systems. However, if the electron is characterized by its Compton radius, $R_C = \hbar/mc$, then h is also identified with the electron – the discrete quantum of matter-wave systems.

Historically, the portion of the particle-wave system that is regarded as being essentially quantum mechanical emerged quite differently in the *massless* electromagnetic system than it did in the *massive* particle-wave system. In the classical era of the nineteenth century, electricity and light were regarded as purely wave phenomena, so that the identification of the quantized photon by Planck and Einstein marked the introduction of quantum mechanics, and thus the end of classical electrodynamics as a complete system. The situation with respect to the electron – the quantum of the electron-wave system – was just the reverse. The electron itself was regarded as a purely classical entity, so that the identification of the electron wave by de Broglie marked the introduction of quantum concepts, and thus an end to the completeness of classical electron theories. In the electromagnetic case the *photon* is considered to be the quantum-mechanical phenomenon, whereas in the electron case the *electron wave* is the quantum-mechanical phenomenon. This situation was summarized by Rosenfeld (1973, p. 260), in a discussion of Bohr's complementarity principle:

"It is curious to point out in this connection that the situation between the complementary aspects of light, the wave aspect which is in direct correspondence with the classical description, and the concept of photon which is symbolical, which is only an expression for the exchange of energy and momentum between matter and radiation, is not paralleled, as de Broglie thought, by the two wave and particle aspects of matter, but it is just the other way around. For matter, the aspect which is in correspondence with classical observation is the particle aspect, of course; whereas the wave aspect is a symbolical one, and is the basis for the statistical description of the processes that can occur, and which Bohr called *individual* processes, meaning thereby that they are not divisible, that they are not analyzable in the classical way."

Today we are accustomed to thinking of *quantum systems* as systems in which the constant h appears, so that *classical systems* are systems in which h does not appear. But, as we discuss briefly in this chapter, the field of *stochastic electrodynamics (SED)* has blurred this distinction. In SED, we find h emerging from some patently non-quantal systems. Hence the classical and quantum mechanical domains of physics appear to be linked together -- as indeed they must be. This implies, in particular, that classical concepts may still have a certain validity deep inside the microworld.

Classical electrodynamics (CED) is characterized by Maxwell's equations, which map out the E and H (electric and magnetic) fields that occupy a region of space. Implicit in these equations is the assumption of a field-free "empty" vacuum state. Thus if the boundary conditions are such that E and H are everywhere equal to zero, then no electromagnetic forces are present. This is the classical viewpoint, and it remained largely unchallenged until 1948. In that year the Dutch physicist Hendrik Casimir proposed an experiment to test the emptiness of the vacuum state. If two electrically-conducting, parallel plates are placed a short distance apart and then charged, a Coulomb force will be produced between the plates. If the charge is removed, then the electric field vanishes, and the force between the plates should drop to zero. But Casimir made a theoretical analysis of the case where the two plates are considered to be immersed in a residual background field of fluctuating electromagnetic radiation, and he concluded that these residual random fields would lead to a net attractive force on the plates. The actual experiment was carried out in 1958 by M. J. Sparnaay, and it yielded the crucial result that when the temperature of the system is lowered towards absolute zero, the force does not vanish, as we would expect from conventional classical considerations, but instead approaches a constant non-zero value (Sparnaay, 1958). This *Casimir* force in dynes is[1]

$$F = -\frac{\pi hcA}{480d^4} , \qquad (4.1)$$

where A is the area of a plate, d is the distance between the plates, and the units are in cm. The force F scales inversely as the fourth power of the separation distance, and the magnitude of the force is determined by Planck's constant h. Thus we have the remarkable result that Planck's constant emerges from what is in essence a purely classical system.

The Casimir effect demonstrates that the vacuum state is filled with a background of zero-point electromagnetic radiation.[2] The spectral shape of this zero-point radiation is[1]

$$\rho(\omega) = \left(\frac{\omega^2}{\pi^2 c^3}\right) \cdot \frac{1}{2}\hbar\omega\,, \tag{4.2}$$

where ω is the radiation frequency. Thus the frequency dependence of the energy density is

$$\rho(\omega) \sim \omega^3\,. \tag{4.3}$$

Eqs. (4.2) and (4.3) have some special properties:[1]

(1) Eq. (4.2) is the expression that follows from *quantum electrodynamics, QED*, wherein we assign a zero-point energy $\hbar\omega/2$ to each oscillator frequency ω.

(2) The cubic frequency dependence shown in Eq. (4.3) is required in order to obtain a Casimir force that varies inversely as the fourth power of the separation distance.

(3) This cubic dependence is also mandated in that it uniquely leads to a Lorentz-invariant background spectrum.

(4) The cubic spectrum remains unchanged under an adiabatic compression, so that the effects of the zero-point radiation are not manifested in "ordinary" experiments of the piston and cylinder type.

Hence there are several reasons that point to a zero-point energy spectrum of the form shown in Eq. (4.2), and in particular to the cubic intensity law of Eq. (4.3). However, since the vacuum-state radiation frequencies ω extend in principle upward without limit, these equations predict an infinite energy density for the vacuum. In actual calculations, only changes or differences in energy are considered, and these invariably turn out to be finite.[2] Thus this energy divergence of the vacuum state poses no serious calculational problems. But it does pose problems for general relativity, and also for the general feelings of physicists, who would prefer to work with mathematically-bounded equations.

When we add this zero-point electromagnetic radiation to classical electromagnetic theory, we have the formalism of *stochastic electrodynamics, SED*.[1] Planck's constant h enters this formalism only through the way in which it normalizes the zero-point radiation. SED is similar to classical electrodynamics in that it features only commuting operators (c-numbers), in contrast to the non-commuting operators (q-numbers) of quantum electrodynamics. But SED features the constant h, which is the hallmark of quantum theories. Thus SED occupies a position that is *between* CED and QED. Noting this situation, Boyer (1980, p. 51) commented as follows:

"... stochastic electrodynamics is like traditional classical electron theory in having the same concepts of particle, force, and field, although it contains Planck's constant h. In the limit $h \to 0$, corresponding to a vanishing spectrum of zero-point radiation, we recover ... traditional classical electron theory ... .

"On the other hand, since stochastic electrodynamics contains Planck's constant *h*, it is in a sense similar to quantum electrodynamics. However, the stochastic theory is a classical theory and has none of the noncommuting operators on Hilbert space which appear in the quantum theory.

"Thus stochastic electrodynamics stands in an intermediate position between traditional classical electron theory and classical (quantum) electrodynamics. Exploration of stochastic electrodynamics should extend our ability to explain atomic phenomena in classical terms and **it should clarify the limits of applicability of a purely classical theory.**" (Emphasis added.)

There is one other remarkable result that emerges from these SED studies. When an observer in an accelerating frame of reference measures the zero-point background radiation, he obtains an enhanced spectrum relative to the one measured by a Galilean observer. If we decompose this enhanced spectrum into the original zero-point spectrum plus an "acceleration component", the acceleration component is seen to have the shape of a Planck thermal radiation spectrum.[2] This result was first obtained by Davies (1975) and by Unruh (1976), and its significance is that the Planck radiation spectrum, which was initially derived by Planck on the basis of quantum-mechanical ideas, can in fact be obtained within an entirely classical context (Boyer, 1984).[3]

There was another hope for Casimir forces, which unfortunately did not turn out to be realized. In addition to calculating the zero-point forces between parallel plates, Casimir also evaluated the case of a spherical conducting shell. In this calculation, the resultant force contained an unspecified constant (de la Peña, 1983, p. 469). In the parallel-plate case, the force was attractive, as shown in Eq. (4.1). Assuming that the force would also be attractive in the spherical-shell case, Casimir proposed that this background-radiation force might serve as the Poincaré stress that holds the electron together (see Chapter 3). However, in a long and very involved calculation, Boyer (1968, 1969) demonstrated that the force is actually repulsive. He stated this conclusion as follows (Boyer, 1968, p. 1765):

"... it seems most melancholy to report that ... the contribution of the zero-point energy ... is of the opposite sign from that suggested. Thus instead of balancing the electrostatic repulsion, the quantum zero-point force also expands the sphere. Our calculation invalidates the Casimir model in the form given here."

This result probably supplied the final *coup de grace* for the concept of a purely electromagnetic electron, since no self-confining mechanism is now to be found among all of the known electromagnetic phenomena.

The concept of stochastic electrodynamics can actually be traced back to an early suggestion of Nernst (1916) that zero-point radiation should be added to the electromagnetic field. The work of Casimir (1948), Sparnaay (1958) and others has demonstrated the practical consequences of this suggestion. However, SED is a difficult subject, and the relatively small number of workers in this area do not always agree among themselves as to exactly what has been accomplished.[4] A detailed description of difficulties that still remain to be dealt with in SED is given by de la Peña (1983, pp. 536-567). He concludes with this assessment (1983, p. 567):

"The picture that emerges is that of a theory with a sound conceptual basis but which is still unfinished. It has achieved a wide range of successful applications, showing that the basic ideas involved cannot be rejected out of hand; it has also come up against a number of important difficulties where it yields unacceptable answers, showing that some significant elements are still lacking."

Thus stochastic electrodynamics, like other areas in physics, still has some frontiers to be explored.

There are two relevant conclusions that we can draw from these studies on stochastic electrodynamics: *(1)* Planck's constant h emerges from an essentially classical treatment of the classical Casimir experiment; *(2)* Planck's radiation law emerges from a classical (general relativistic) treatment of the classical Casimir vacuum state as viewed in an accelerating frame of reference. The *first* conclusion reveals quite clearly that *Planck's constant h* is not unique to what we characteristically think of as quantum systems, but in fact also occurs in the fundamental background component of classical electromagnetic theory. The *second* conclusion reveals that the *Planck radiation spectrum*, which features the constant h, and which is customarily derived by imposing an energy quantization requirement, can also be derived classically without saying anything explicit about energy quantization. Thus h is in fact not unique to q-number quantum systems, but also plays a role in the c-number theories that we have heretofore thought of as constituting the classical domain. Hence the dichotomy of classical and quantum physics stems more from the incompleteness of our understanding than from any natural division that exists in nature. We should expect to find classical and quantum ideas intertwined, so that the boundary of the microworld does *not* represent a jumping-off point in which one goes from exclusively classical to exclusively quantum concepts. We can no longer draw a boundary between classical and quantum systems on the basis of size, and we can no longer draw it on the basis of Planck's constant h.

Notes

[1] For articles that give a survey and critique of the concepts of stochastic electrodynamics, see Boyer (1980, pp. 49-63) and de la Peña (1983, pp. 428-581).

[2] For a recent summary of these ideas, see *"The Classical Vacuum"*, (Boyer, 1985).

[3] Derivations of the Planck radiation spectrum are discussed in de la Peña (1983, pp. 472-477).

[4] See Boyer (1980, p. 49).

Part II

The Natural Size of the Electron

"There is nothing more difficult to take in hand, more perilous to conduct, or more uncertain in its success than to take the lead in the introduction of a new order of things, because the innovator has for enemies all those who have done well under the old condition, and lukewarm defenders in those who may do well under the new."

Niccolo Machiavelli, *Il Principe (1513)*

Chapter 5

The Natural Size of an Elementary Particle

Is there a natural size for an elementary particle? For example, if we know a particle's mass, do we also know its radius? It is apparent from Figure 1.1 that answers to these questions do not yet exist for the electron. The conceptual difficulty in the case of the electron is not just that scattering experiments indicate a small size – but that they indicate no detectable size at all, even down to limits which are way below any of our classical expectations. If we study the muon instead of the electron, we arrive at this same frustrating impasse.

Fortunately, there are four elementary particles whose sizes have been directly measured, and we can use these to answer the above questions. However, somewhat unfortunately, all four of these particles are believed to be compound states, each formed from combinations of quark substates. This makes the analysis more complicated, but we can nevertheless arrive at a fairly definitive conclusion. The four particles with known sizes are the proton, neutron, charged pion, and charged kaon, which have the charge states p^+, n^0, $\pi^\pm$, and $K^\pm$, respectively. In each case, the measured size of the particle is its electromagnetic size – electric and/or magnetic – which is customarily expressed in terms of a root mean squared (rms) radius. The natural length scale for these particles is in units of fermis (10^{-13} cm). The masses of these particles are well known,[1] and their rms electromagnetic radii have been accurately determined experimentally.[2,3] The electrically neutral pion and kaons – π^0, K_S, K_L – do not have measured sizes.

In the cases of the charged pions and kaons, the rms radii of their *charge* distributions have been measured. These are spin zero particles, and they do not possess *magnetic* radii. In the case of the spin one-half proton, both its *charge* radius and its *magnetic* radius have been measured, and these turn out to be essentially equal to one another.[2] This is an important result, since it indicates that *Ampere's hypothesis* (Feynman, 1961, p. 84), which is the assumption that magnetic fields arise from rotating charge distributions, applies to the proton. In the case of the spin one-half neutron, its measured rms *charge* radius is approximately zero, and its measured rms *magnetic* radius is approximately equal to that of the proton.[2] This indicates two things: *(1)* the neutron and proton have the same sizes;

(2) the neutron, although electrically neutral, nevertheless contains internal plus and minus charges. The first point is a result that we expect for the neutron and proton, since they are the two members of an isotopic spin multiplet, and they transform readily into one other via the processes of beta decay and positron decay. The second point, when combined with the first, is another indication of the applicability of Ampere's hypothesis.

The rms electromagnetic particle radii that have been measured[2,3] are summarized in the following table:

TABLE 5.1. EXPERIMENTAL RMS ELECTROMAGNETIC RADII IN FERMIS.

Particle	*Electric radius*	*Magnetic radius*
Proton[2]	0.82 ± 0.02	0.82 ± 0.02
Neutron[2]	~ 0	$\sim 0.82 \pm 0.02$
Pion[3]	0.66 ± 0.01	–
Kaon[3]	0.58 ± 0.04	–

As can be seen in Table 5.1, all of these particles have electromagnetic radii that are in the range of 0.6 - 0.8 fermi. From a theoretical point of view, what radii would we naturally expect to find for them? The one answer that immediately comes to mind is the Compton radius, $R_C = \hbar/mc$. Since the masses of the particles are known,[1] we can use these to define *particle Compton radii*, which we can then compare to the measured rms electromagnetic radii. This comparison is given in Table 5.2.

TABLE 5.2. EXPERIMENTAL RADII AND "PARTICLE COMPTON RADII"

Particle	*Experimental rms radius (fm)*	*Particle mass (MeV)*	*Particle Compton radius (fm)*
Proton	0.82	939	0.21
Neutron	0.82	940	0.21
Pion	0.66	140	1.44
Kaon	0.58	493	0.40

Each of the particle Compton radii shown in Table 5.2 is within a factor of four of the experimental rms radius. Thus the particle Compton radius provides at least an order-of-magnitude estimate of the size of the particle. However, we may be able to do better than this. Since these particles are believed to be formed from quark substates (Feld, 1969), we can try using *quark Compton radii* rather than *particle Compton radii*, and see if this improves the results shown in Table 5.2. But this brings up a difficulty. In order to calculate a quark Compton radius, we need to know the quark mass, and the subject of quark masses continues to be one of the major mysteries in the field of elementary particle physics. We can at least partly obviate this difficulty by using *constituent-quark* masses,[4] although it must be kept in mind that the concept of constituent-quark masses is not very clear-cut.[5] Constituent-quark masses are quark masses whose sum adds up to roughly the observed mass of the particle. In the standard quark model, the particles shown in Table 5.2 are constructed from three types of quarks: *up* quarks u, *down* quarks d, and *strange* quarks s, together with their corresponding antiquark states. The proton contains two u quarks and one d quark, and the neutron contains one u quark and two d quarks. Thus, from the constituent-quark viewpoint, each u and d quark will have approximately one-third the mass of a proton or neutron, or about 313 MeV. The pion is constructed from one u or d quark and one $\overline{u}$ or $\overline{d}$ antiquark.[6] The kaon (antikaon) is constructed from one u or d quark (one $\overline{u}$ or $\overline{d}$ antiquark) and one $\overline{s}$ antiquark (one s quark). The s quark is commonly assigned a constituent-quark mass which is roughly 150 MeV larger than the u and d masses, or about 500 MeV.[7] If we use these constituent-quark mass values to calculate *quark Compton radii*, we obtain the results that are displayed in Table 5.3:

TABLE 5.3. EXPERIMENTAL RADII AND "QUARK COMPTON RADII".

Particle	*Experimental rms radius (fm)*	*Quark substructure*	*Quark Compton radii (fm)*
Proton	0.82	uud	0.6, 0.6, 0.6
Neutron	0.82	udd	0.6, 0.6, 0.6
Pion	0.66	$(u\overline{d})$ or $(\overline{u}d)$	0.6, 0.6
Kaon	0.58	$(u\overline{s})$ or $(\overline{u}s)$	0.6, 0.4

As can be seen in Table 5.3, the agreement between the experimental particle radii and the *quark Compton radii* is much closer than was the agreement with the *particle Compton radii* shown in Table 5.2. In particular, a cluster of three u and d quarks, each with a radius of 0.6 fermi, leads to a calculated overall rms radius of just about 0.8 fermi for the proton or neutron, in good agreement with the experimental value of 0.82 fermi. And the pion and kaon size relationships seem equally accurate. This suggests that the size of a composite particle is derived primarily from the sizes of its constituent quarks. Hence we should assess the Compton nature of a *composite* particle in terms of the Compton radii of its *constituent-quark substates.* When we do this, we obtain the following significant conclusion:

> The particles shown in Tables 5.1 - 5.3, which are *the only elementary particles whose sizes have been directly measured,* all have *Compton-like geometries,* with an overall accuracy that can be reasonably estimated as ± 30%. (5.1)

In addition to the direct size measurements displayed in Table 5.1, we can also obtain some information as to the sizes of the proton and neutron on the basis of their spins and magnetic moments. To accomplish this, we assume that each of these particles is composed of three constituent quarks, and we represent each quark as a relativistically spinning sphere, using the sphere model that is described in detail in Part III. If each quark is assigned a constituent-quark Compton radius, as displayed in Table 5.3, then the *calculated* quark spin angular momentum turns out to be $1/2\,\hbar$, as is required phenomenologically in

SU(6). Furthermore, if each quark is assigned an equatorial point charge, $q = +2/3\ e$ for the *u* quark and $q = -1/3\ e$ for the *d* quark, then the quarks have *calculated* magnetic moments that are in agreement with the postulates of SU(6) (Feld, 1969, p. 339). Thus we have the important result that *Compton-sized u* and *d* quarks accurately reproduce both the *spins* and the *magnetic moments* of the proton and neutron, as well as combining together to closely match the *over-all rms radii* of these particles.

The above results can be extended, although somewhat indirectly, up to the mass region of the charmed quark, *c*. The *u* and *d* quarks we have just discussed have Dirac-like magnetic moments, $\mu = q\hbar/2mc$, where m is the constituent mass of a quark and q is the charge on the quark. The magnetic moment of the *c* quark cannot be measured directly, but it can be deduced on the basis of the magnetic dipole transitions that occur between the atomic-like energy levels of the charmonium system.[1] Each of these magnetic transitions corresponds to the spin flip of *one* quark, and hence is proportional to the magnetic moment of the quark. The experimental charmonium transition rates are consistent with the assumption of a Dirac magnetic moment for the *c* quark (Gottfried and Weisskopf, 1984, p. 112). Thus, if we apply the relativistically spinning sphere model to this situation, we are led to a Compton-sized radius for the *c* quark, which is known to have a mass of roughly 1500 MeV.

We can summarize this discussion as follows. The only elementary particles whose sizes are directly known – the proton, neutron, charged pions, and charged kaons – all have spatial dimensions which are roughly comparable (within a factor of four) to their *particle Compton radii,* as shown in Table 5.2. Furthermore, their spatial dimensions are even more accurately delineated (to within ± 30%) by the *quark Compton radii* of their constituent-quark substates, as shown in Table 5.3. Also, in the cases of the proton and neutron, the spatial dimensions of their constituent quarks are consistent with the SU(6) values for the spins and magnetic moments of these quarks, as calculated via the systematics given in Part III. And there is at least indirect evidence that this mass versus particle relationship extends up to the mass region of the charmed quark. Thus all of the available experimental evidence suggests that *there is a natural size for an elementary particle* – at least for one that is composed of quarks:

> *The natural size of an elementary particle is quite accurately given as the cluster size of its constituent quarks, where each constituent quark with mass m_Q is assigned the constituent-quark Compton radius, $R_Q = \hbar/m_Q c$.* (5.2)

Now, how does all of this apply to the electron? Since the electron does not seem to have quark substates, it is not immediately clear as to how it is related to conclusion (5.2). But from the results we develop in Part III of the monograph, it will become apparent that the main properties of the electron can be obtained by representing it as a relativistically spinning sphere – that is, as a single quark-like state. Hence the electron emerges phenomenologically as a quark-like object, except for the facts that *(a)* it carries an integral rather than a fractional charge, and *(b)* it interacts only electromagnetically and not hadronically. If we knew nothing experimentally about the size of the electron, and had only the considerations of the present chapter to draw on, it would seem reasonable to assume that the size of the electron is comparable to its Compton radius. This is a legacy that we have from particle physics, and it should not be lightly disregarded.

Notes

1 See Particle Data Group (1992).

2 See Gottfried and Weisskopf (1986, pp. 279-281).

3 See Amendolia *et al.* (1984, 1986a, 1986b).

4 Constituent-quark masses are assumed to be mainly responsible for the observed mass of a particle. Constituent-quark binding energies are taken to be small, except for the very lightest composite meson states.

5 The question of quark masses is still very much up in the air. The *constituent-quark* masses described here are phenomenologically useful in reproducing the observed masses of baryons and mesons. However, present-day theoretical approaches tend to emphasize *current-quark* masses, which are the masses that one would, at least in principle (but not yet in practice), use in the interaction Lagrangian, and which are very much smaller than constituent-quark masses (the u and d current-quark masses are approximately equal to zero).

6 If we assign the u and d quarks constituent-quark masses of 313 MeV, then we can quite accurately reproduce the masses of the baryon octet by using zero binding energies. However, the binding energy of the pion, which is also constructed of u and d quarks, then becomes enormous. But since baryons feature only quark triplets, whereas mesons feature quark-antiquark pairs, mesons may in fact have much larger binding energies than baryons.

7 In both the constituent-quark approach and the current-quark approach, the s quark is assumed to be roughly 150 MeV more massive than the u and d quarks. An s quark mass of about 500 MeV gives a reasonable fit to the baryon octet states, but it leads to a very large constituent-quark binding energy for the kaon.

Chapter 6

The Spectroscopic and Bulk Sizes of the Electron

Broadly speaking, there are three methods we can utilize in deducing the size of an elementary particle: *(1)* we can make a direct measurement of its size, which is usually its electromagnetic size, and which is generally obtained via scattering experiments; *(2)* we can use theoretical models to relate its known spectroscopic properties to the required sizes that are needed in order to account for these properties; *(3)* we can determine its average size as manifested in a close-packed aggregate of particles. We invoked the first method in Chapter 5, where we discussed the only elementary particles whose sizes have been directly measured – the proton, neutron, pion, and kaon. Unfortunately, these particles do not include the electron, which has thus far defied all efforts at pinning down a measurable size. In the present chapter we take up the second and third methods and apply them to the electron. The second method – the use of spectroscopic information and theoretical models – is treated here in a non-relativistic context. (In Part III we give a relativistic treatment of this problem.) The third method – that of bulk measurements – is obtained by a somewhat speculative use of astrophysical data. In Chapter 7 we return to the first method, and we discuss the experiments that have a bearing on the size of the electron. These experiments primarily involve the *electrostatic* properties of the electron, and they lead to extensive ramifications with respect to the nature of the mass of the electron. Then in Chapter 8 we study the *magnetic* properties of the electron, using mainly the second method: namely, an examination of the magnetic spectroscopy of the electron. The magnetic anomaly in the electron – its anomalous magnetic moment – can be related to the mass components of the electron in a very intriguing manner. The size information that we obtain from each of these studies is somewhat indirect, but it is informative. The conclusion we reach in Part II is that these results combine together to form a cohesive mosaic of electron properties.

A. The Spectroscopic Size of the Electron

Suppose that a classical physicist in the year 1900 were to be supplied with the following properties of the electron, and then asked to deduce its size:

mass	$m = 0.511\ \mathrm{MeV/c^2}$
electric charge	e = unit charge in esu
spin angular momentum	$J = 1/2\,\hbar$
magnetic moment	$\mu = e\hbar/2mc$

He would have as theoretical tools the equations of non-relativistic Newtonian mechanics and the equations of classical electrodynamics. Now, it is possible to obtain some size estimates of the electron by assuming that its *mass* is electromagnetic, and this possibility is investigated in Chapters 7 and 8. But here we simply assume that the mass is an input parameter, and is handled as any mass would be in the macroscopic world. Thus our classical physicist must base his size estimates on the dynamical characteristics of the electron – its spin and magnetic moment.

We start with mechanical considerations, as embodied in the *spin angular momentum* of the electron. Angular momentum $\vec{J}$ is defined classically by the equations

$$\vec{J} = I\vec{\omega} \tag{6.1}$$

and

$$I = \sum_i m_i r_i^2 , \tag{6.2}$$

where I is the moment of inertia with respect to the axis of rotation, ω is the angular velocity of the rotation, m_i is the *ith* mass element, and r_i is the distance of the *ith* mass element from the rotational axis. In the case of a uniform sphere of matter of radius R, which we can take as the prototype for an electron, the non-relativistic moment of inertia is

$$I_{\text{non-rel}} = \frac{2}{5} mR^2 . \tag{6.3}$$

The angular velocity is $\omega = \mathrm{v}/R$, where v is the peripheral linear velocity. These values give

$$J = \frac{2}{5} mR\mathrm{v} . \tag{6.4}$$

If we now set J equal to the spin $1/2\,\hbar$ of the electron, we obtain the following equation for R:

$$R = \frac{5}{4}\left(\frac{\hbar}{m\mathrm{v}}\right) . \tag{6.5}$$

Looking ahead a few years to the introduction of relativity, we can have our classical physicist rewrite this equation as

$$R = \frac{5}{4} R_C \left(\frac{c}{\mathrm{v}}\right) , \tag{6.6}$$

where $R_C = \hbar/mc$ is the Compton radius of the electron, and c is the velocity of light. Thus if the peripheral velocity v of the electron is comparable to the velocity of light c, then the radius of the electron is comparable to its Compton radius R_C. If this is in fact the case, then the electron is following the scaling law for sizes that we deduced in Chapter 5 on the basis of the experimental size measurements of the proton, neutron, pion, and kaon, all of which have Compton-like geometries.

Now consider the problem of the size of the electron from the standpoint of classical electromagnetic theory. What does the *magnetic moment* of the electron tell us about its size? In classical E & M, a magnetic field arises from the motion of an electric charge. Specifically, a magnetic dipole moment, such as is observed in the electron, corresponds to a current loop. Using c.g.s. units, we have the current-loop equation

$$\mu = \pi R^2 \cdot i , \tag{6.7}$$

where μ is the magnetic moment, R is the radius of the current loop, and i is the current flowing in the loop. If we attribute this current to the rotation of a point electric charge e, then

$$i = \left(\frac{e}{c}\right) \cdot \left(\frac{\omega}{2\pi}\right) = \left(\frac{e}{c}\right) \cdot \left(\frac{\mathrm{v}}{2\pi R}\right) , \tag{6.8}$$

where $\omega = \mathrm{v}/R$ is the angular velocity of the charge e, and v is the instantaneous linear velocity. Inserting Eq. (6.8) into Eq. (6.7), we obtain

$$\mu = \left(\frac{eR}{2}\right) \cdot \left(\frac{\mathrm{v}}{c}\right) \tag{6.9}$$

as the theoretical expression for the magnetic moment μ. Experimentally, the first-order magnetic moment of the electron is

$$\mu = \frac{e\hbar}{2mc} . \tag{6.10}$$

Eliminating μ from Eqs. (6.9) and (6.10), we obtain the following expression for R:

$$R = R_C \left(\frac{c}{\mathrm{v}}\right) . \tag{6.11}$$

Apart from a factor of 5/4, Eq. (6.11) is identical to Eq. (6.6). Thus the *classical spin angular momentum* and the *classical magnetic moment* of the electron both point to essentially the same size for the electron.

We can sharpen this analysis considerably if we now convert our year 1900 physicist into a 1905 physicist, and allow him to use the newly discovered tools of special relativity. Considering the possibility that the peripheral linear velocity v of the spinning sphere may indeed be comparable to the velocity c, he would realize that the moment of inertia I in Eq. (6.3) should be calculated relativistically. If he carried out this calculation in a straightforward manner, as we do in Chapter 10, he would first of all discover the non-trivial fact that the calculated relativistic moment of inertia I_{rel} is *finite*. Then he would go on to obtain the quantitative result that

$$I_{rel} = \frac{1}{2} m R^2 , \tag{6.12}$$

where m is now the (observed) spinning mass (rather than the stationary rest mass) of the electron. If we use the relativistic moment of inertia I_{rel} (Eq. 6.12) rather than the non-relativistic moment of inertia $I_{non\text{-}rel}$ (Eq. 6.3), then Eq. (6.6) becomes

$$R = R_C \left(\frac{c}{\mathrm{v}} \right), \tag{6.6$'$}$$

which is identical to Eq. (6.11). The annoying factor of 5/4 has disappeared. Hence the spin angular momentum and the magnetic moment of the electron, which are the only properties that give us direct information about its radius, both indicate *precisely* the same size for the electron. This is a result that should not be cavalierly ignored. Moreover, special relativity gives us an even more important piece of information. Before the advent of special relativity, we had no basis for concluding anything about the instantaneous peripheral velocity v of the spinning electron. But special relativity makes a definitive statement here. It tells us that the velocity v cannot exceed the velocity c. Thus Eq. (6.6′) for the spin angular momentum and Eq. (6.11) for the magnetic moment both become lower bounds. They both yield the equation

$$R \geq R_C , \tag{6.13}$$

which states that the radius R of the electron is equal to or greater than the electron Compton radius R_C. Our classical physicist could then go on to conclude from this result that the mass of the electron does not arise from electrostatic self-energy (Chapter 7), and he might even conclude something about the anomalous magnetic moment of the electron (Chapter 8). If we didn't tell him about the experiments that indicate point-like scattering, he would feel quite comfortable with the concept of a Compton-sized electron.

B. Bulk Measurements of Particle Sizes

If we have a group of elementary particles clustered closely together, we can use the average volume occupied by each particle as at least a rough measure of the size of the particle. However, the attainment of this close clustering requires a very powerful attractive force. Two types of systems in which this clustering occurs are *stellar bodies*, where long-ranged gravity provides the strong attractive force, and *atomic nuclei*, where short-ranged nucleon-nucleon interactions provide the strong attractive force. In these systems, the attractive force is counterbalanced by the structural integrity of the particles that make up the system. That is, the particles are compressed together but not crushed by the attractive force. In the case of atomic nuclei, the structural integrity is provided by the stability of the constituent protons and neutrons. This stability is conventionally characterized by the form of the nucleon potential – namely, the so-called nucleon *hard core potential* – which is attractive at distances greater than about a fermi, but which turns strongly repulsive at shorter distances. The situation with respect to the stability of stellar systems is considerably more complex, as we now discuss.

In stellar bodies, the structural integrity occurs on *four* different levels, which are governed by the total mass and the energy content of the star. A large mass leads to a large overall gravitational force. But a large energy content corresponds to large kinetic energies, which serve as an effective buffer against the compressive gravitational forces. In relatively young, energetic, and modestly sized stellar bodies such as the *Sun*, the atomic states are highly ionized, but typical atomic electron spacings are still maintained on the average, and it is the overall size of an atom that sets the scale for the stellar dimensions. This condition also prevails in small cold bodies such as the *Earth*, where kinetic energies are small, but where the overall gravitational force is not sufficient to crush the constituent atoms that make up the body. This is the first level of stellar structural integrity. The second level of stellar integrity is found in *white dwarf stars*. These older stars have exhausted so much of their energy that the atomic structures can no longer be maintained against the crush of gravity. In these stars, the electron now takes over as the stabilizing element, and electron pressure counterbalances the gravitational forces. The third level of stellar integrity occurs in *neutron stars*. These somewhat more massive stars have so little kinetic energy and such strong gravitational forces that even the electron pressure is overcome, and the electrons and protons are squeezed together to form an aggregate "crystal" of neutrons. The stabilizing force in neutron stars is provided by the structural integrity of the neutron. The fourth and final level of stellar integrity occurs in the very massive *black holes*. These stellar bodies have such tremendous gravitational forces that even nucleon structure cannot be maintained, and the star collapses into what is in essence a "space-time point". The integrity of a black hole is provided by the integrity of the space-time continuum itself. Since photons cannot escape from the powerful gravitational pull of a black hole, direct observational evidence of these objects is lacking. Hence the name "black hole".

The information about particle sizes that we can extract from this discussion is summarized in Table 6.1, which uses Eq. (6.14) to relate bulk densities to effective particle radii:

TABLE 6.1. ESTIMATION OF PARTICLE SIZES FROM BULK DENSITIES.

Aggregate system	*Density range in g/cm³*	*Structural particle*	*Particle radius in cm (Eq. 6.14)*
Normal star	10^{-4} to 10	hydrogen atom	$(0.3-13)\times10^{-8}$
White dwarf	10^4 to 10^8	electron	$(1.3-28)\times10^{-11}$
Neutron star	10^{11} to 10^{15}	neutron	$(0.6-13)\times10^{-13}$
Atomic nucleus	2×10^{14}	nucleon	1.0×10^{-13}

This table gives density ranges for normal, white dwarf, and neutron stars (Ostriker, 1971), together with the nucleon (proton plus neutron) density that occurs in a large atomic nucleus (Cameron, 1971, p. 327). If we relate the densities ρ to the dimensions of the basic structural particles by means of the equation

$$\rho = \frac{m}{(2r)^3}, \qquad (6.14)$$

where *m* is the particle mass and *r* is the radius, we obtain the inferred particle radii shown in the last column of Table 6.1. The *nucleon* radius of one fermi that is deduced in this manner in Table 6.1 agrees well with the proton and neutron radii that are obtained by direct measurements (Table 5.1). It is also in general agreement with the *neutron* radius that we infer in Table 6.1 from the systematics of a neutron star. Similarly, the radius of a *hydrogen atom* (0.5×10^{-8} cm) is in broad agreement with the radius that is deduced from the density of a normal star. (It is in exact agreement in the case of the Sun!) As a final result, the *electron* radius that we deduce in this manner from a white dwarf star turns out to be *Compton-sized*, $\sim 4 \times 10^{-11}$ cm, which is the whole point of the present discussion. The conventional explanation for the observed electron spacing in a white dwarf star is in terms of degenerate electron pressure (Schwarzschild, 1958, p. 56), wherein the operation of the Pauli exclusion principle on the available electron phase space forces the electrons into states with very high momentum values. However, the electron systematics displayed in Table 6.1 may possibly also be a consequence of the basic size of the electron itself. The evidence here is not overwhelming, but, when combined with the other information we have about the size of the electron, it is at least suggestive.

Chapter 7

The Electric Sizes and Electric Self-Energy of the Electron

The dominant feature of the electron is its electrostatic charge e. One of the important endeavors in twentieth century particle physics has been the attempt to determine the spatial extent of this charge. There are actually two sizes that we are forced to consider. The first is the *intrinsic size* of the charge itself, which we characterize as the *electric charge radius* R_E (Eq. 1.1d). The second is the *spatial distribution* of the charge, as it is manifested in various experiments that involve electrons. We can characterize this spatial distribution in terms of a *charge distribution radius* R_{CD}. Thus we have

$$R_E \equiv \textit{intrinsic size of the charge } e\ ; \tag{7.1}$$

$$R_{CD} \equiv \textit{charge distribution of the electron, as manifested experimentally}\ . \tag{7.2}$$

The radii R_E and R_{CD} are not equivalent. For example, the experimental value that emerges for R_{CD} is radically different in electron scattering experiments than it is in Lamb shift experiments, as we have already discussed. Operationally, the only charge size we can measure directly is R_{CD}. The radius R_E is an inferred quantity. We will assume that R_E is an *invariant*, so that *it does not change from experiment to experiment*. The *smallest* measured value of R_{CD} thus constitutes an *experimental upper bound* for the value of R_E. As we discuss later on in this chapter, one of the best ways of gaining information as to the intrinsic nature of the charge e may be in terms of its effect on the electrostatic self-energy W_E of the electron.

If we attempt to measure the size of the charge on the electron, one of the first things we must determine is the *shape* that we attribute to this charge. When the only things we knew about the electron were its charge e and mass m, the natural assumption to make was that the charge is *spherically symmetric*, and this assumption was featured in all of the early calculations of electron structure. When we later learned about the spin S and magnetic moment μ of the electron, a second choice opened up – a charge distribution that is *axially symmetric*. However, both of these choices lead to observational difficulties: namely, if we apply classical electromagnetic theory to each of these charge distributions, there are reasons why their true *electric* sizes may be concealed experimentally. Consider first the case of a *spherically symmetric charge distribution*, which is usually taken to be a solid sphere or a spherical shell. The electromagnetic villain that arises here is *Gauss's Law*, which can be written in integral form as (Jackson, 1962, p. 6)

$$\oint_S \vec{E} \cdot \vec{n} ds = 4\pi \int_V \rho(\mathbf{x}) d^3\mathbf{x} = 4\pi q \ , \tag{7.3}$$

where S is a spherical surface that surrounds the charge distribution $\rho(\mathbf{x})$, where $\vec{E} \cdot \vec{n} \equiv E_N$ is the electric field that emerges normal to the surface S, and where q is the total charge contained in the volume V. If $\rho(\mathbf{x})$ is spherically symmetric, then the electric field E_N is isotropic, and it corresponds to the field produced by a charge q located at the center of the volume V. Thus, to an *external* observer, the extended charge $\rho(\mathbf{x})$ appears to be a point charge q located at the center. An *internal* observer also deduces that the charge is point-like, but he concludes that it has a reduced value $q' < q$.[1] Hence, in order to reveal the distributed nature of the spherically symmetric charge $\rho(\mathbf{x})$, it is necessary for the measuring probe to penetrate into the *interior* of the charge region V, so that it senses a decrease in the effective Coulomb potential. In the case of an electron bound into a P-state atomic orbital, for example, the electron never reaches the position of the nucleus. Thus if its charge distribution is spherical, the electron will exhibit the Coulomb potential of a point charge, regardless of its actual size (which is assumed to be small as compared to orbital dimensions).

Now consider the observational problems that arise with respect to an *axially symmetric charge distribution*. If the electron has a finite size, and if it is rotating, as its spin angular momentum $J = 1/2\,\hbar$ mandates, then its charge distribution is also rotating. This rotating charge distribution corresponds to a set of axially centered current loops. In the limiting case of an equatorial point charge e on the electron, the charge distribution is in the form of a single current loop, which is the case we are interested in. This is where the observational difficulties appear, as we will soon see. A current loop gives rise to a magnetic moment, in agreement with Ampere's hypothesis. This magnetic moment is considered in detail in later chapters. In the present chapter we are only concerned with the *electrostatic potential* that is associated with the current loop. This electrostatic potential leads to spherically asymmetric Coulomb forces which serve to delineate the size of the current loop. In particular, if the electrostatic field that surrounds a current loop is different from the electrostatic field that surrounds an equivalent point charge positioned at the center of the loop, we can use this difference to deduce the spatial extent of the loop, as we now describe.

Figure 7.1 shows a rotating current loop. The spin axis ω forms angles θ and ψ with respect to the z and x axes, respectively, and the projection of ω onto the x,y plane forms a precession angle φ with respect to the x axis. The z axis serves as the quantum-mechanical spin quantization axis, and the spin vector ω precesses around the z axis. The electrostatic potential V of the current loop at points along the z axis is given by the equation (Smythe, 1939, p. 138)

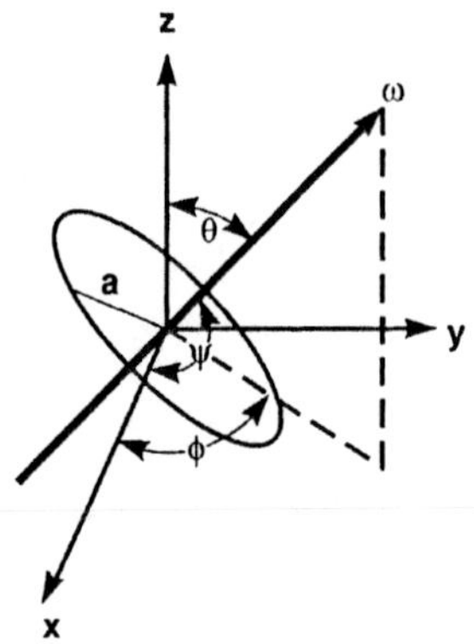

Fig. 7.1. A current loop, as projected onto x, y and z axes. When the angle $\theta = \pm\,\theta_{QM}$ (Eq. 7.8), this is referred to as a *quantum current loop*.

$$V = q/a \sum_{n=0}^{\infty} (-1)^n \frac{1.3\ldots(2n-1)}{2.4\ldots 2n} (a/r)^{2n+1} P_{2n}(\cos\theta) \ , \tag{7.4}$$

where a is the radius of the ring, r is the distance from the center of the ring to the point on the z axis, and $r > a$. The first term in this series,

$$V_0 = q/r \ , \tag{7.5}$$

is the *Coulomb potential of a point charge*. The second term,

$$V_2 = -\frac{1}{2}(q/r)(a/r)^2 P_2(\cos\theta) \ , \tag{7.6}$$

is the *electric quadrupole moment* of the current loop, and is the term of interest here. Since we have $a << r$ in practical applications, it is mainly this term that causes appreciable deviations from a point potential. Now, the Legendre function P_2 is

$$P_2(\cos\theta) \equiv \frac{1}{2}(3\cos^2\theta - 1) \ , \tag{7.7}$$

so that the electric quadrupole moment vanishes at the angles $\pm\,\theta_{QM}$, where

$$\theta_{QM} = \arccos(1/\sqrt{3}) \ . \tag{7.8}$$

This is an extremely interesting result. The reason for this is that $\pm\,\theta_{QM}$ are the quantum-mechanical angles (as we have indicated by the subscript) which project the total spin vector $J = \sqrt{1/2\,(1/2+1)}\,\hbar$ of a spin 1/2 particle onto the z axis with the quantum-mechanically prescribed values $J_Z = \pm\,1/2\,\hbar$. And the electron, of course, is a spin 1/2 particle. Thus, according to the vector rules of quantum mechanics, the current loop that constitutes the rotating charge on the electron is inclined at an angle $\pm\,\theta_{QM}$ that produces a *vanishing electric quadrupole moment with respect to the z-axis of quantization.*[2]

This same result carries over, but in a somewhat different manner, to the x and y axes. The electrostatic potential V_2 at a point on the x axis is proportional to $P_2(\cos\psi)$. Using the relationship $\cos\psi = \sin\theta_{QM}\cos\phi$, and inserting $\sin\theta_{QM} = \sqrt{2/3}$, we obtain

$$P_2(\cos\psi) = \frac{1}{2}(2\cos^2\phi - 1) \ . \tag{7.9}$$

Now, the average value of $\cos^2\phi$ over one cycle of revolution of the angle ϕ is 1/2, so that the expression on the right side of Eq. (7.9) averages out to zero over a cycle of precessional motion. Hence the electric quadrupole moment of the current loop vanishes *on the average* along the x axis, and by symmetry also along the y axis.

We will denote an equatorial current loop that is oriented with respect to the z axis at a quantum-mechanical angle $\pm\theta_{QM}$ as a *quantum current loop*. We have thus established the following crucial result:

> *The electric quadrupole moment of a* ***quantum current loop*** *vanishes identically along the z-axis of quantization, and it also vanishes along the x and y axes when averaged over a cycle of precessional motion.* (7.10)

Hence the deviations from a point charge electrostatic potential for a quantum current loop arise mainly from a higher multipole moment of the loop, the third term in Eq. (7.4):

$$V_4 = \frac{3}{8}(q/r)(a/r)^4 P_4(\cos\theta) \ . \tag{7.11}$$

To evaluate the magnitude of this V_4 multipole effect in a typical atomic case, consider the electrostatic potential of a Compton-sized current loop of radius $a = \hbar/mc$, which has an electric charge $q = e$, and which is located at a distance r from the nucleus, where r is equal to the first Bohr orbit in hydrogen – $r = \hbar^2/e^2m$ – with m the electron mass. Under these assumptions, we obtain $a/r = e^2/\hbar c = \alpha \cong 1/137$. Thus, in an atom, the contribution of the electric multipole moment V_4 of the current loop is smaller than the point charge term V_0 by a factor which is of order α^4 in the fine structure constant α.

The problems that arise in trying to deduce the spatial extent of a spherically symmetric charge distribution have long been recognized (Rohrlich, 1965, pp. 127-128). But the similar problems that apply in the case of an axially symmetric quantum current loop do not seem to have been pointed out except by the present author (Mac Gregor, 1978, pp. 79-81).

The question as to the size of the electric charge on the electron is of dominant interest not only for the information it provides with respect to the actual size of the electron itself, but also for the information it suggests with respect to the nature of the mass of the electron. Specifically, how much of the total mass of the electron is attributable to the self-energy W_E of the charge *e?* To pinpoint this discussion, let us pose two questions:

What is the intrinsic size R_E of the charge on the electron? (7.12)

Does this charge contribute to the mass of the electron? (7.13)

Now let us recast these questions by bringing out their operational aspects:

How far down in size does the charge e on the electron continue to exhibit a $1/r^2$ Coulomb force law? (7.12*a*)

Does the electrostatic charge distribution $\rho(\mathbf{r})$ of the charge e interact with itself? (7.13*a*)

Let us answer question (7.12*a*) first, because the experimental answer to this question has a profound effect on possible answers to question (7.13*a*). Electrons, being leptons, interact only electromagnetically with other particles. Since electric forces are in general much stronger than magnetic forces, the primary force exerted by an electron is electrostatic in nature. If we investigate this electric force by using an *external* charged probe, then Gauss's Law (Eq. 7.3) guarantees that, if the charge is spherically symmetric, we will find just the Coulomb force,

$$F = \frac{eq}{r^2}, \tag{7.14}$$

where e is the charge on the electron, q is the charge on the probe, and r is the separation distance between the charges e and q. In order to find a force that falls off slower than $1/r^2$, the probe must penetrate into the *interior* of the charge distribution of the electron. Since the electron appears to be a very small particle, the smallest probe we can think of using is another electron, or, equivalently, a positron. Thus electron-electron or electron-positron scattering becomes the definitive type of experiment that we invoke in order to investigate the size of the electron, or, more precisely, the size R_E of the charge on the electron.

In electron-electron scattering, the form of the angular distribution of the scattering events reveals the force law that is operating to cause the scattering. An incoming electron has forward inertia which tends to keep it moving forward. In order to produce a large lateral deflection, a large lateral impulse must be imparted to the electron at the moment of impact. If we know the overall kinematical factors that are present in the scattering process, we can calculate the force that is required in order to produce a certain scattering angle. This force depends on the distance between the two particles at the time the scattering occurs, which is customarily characterized by the *impact parameter* for the scattering event – the distance of closest approach that would occur if the interaction force were to be turned off. If we know the distribution of impact parameters in the ensemble of scatterings that constitutes the experiment, we can calculate the distribution of scattering angles, which is the angular distribution. If the force is weaker than expected, then the number of *large-angle* scatterings will be smaller than expected. Hence the way to discover if the electron-electron scattering process has caused the two particles to interpenetrate into the interior of each other's charge distribution, and hence weaken the Coulomb force below the value indicated in Eq. (7.14), is to look for a fall-off of the wide-angle scattering. Since the small impact parameters that lead to interpenetration require large incident energies (in order to overcome the repulsive Coulomb barrier between the electrons), a fall-off of the wide-angle scattering below its expected value should show up in high-energy scattering experiments.

Now, in order to ascertain whether or not the wide angle scattering has fallen below its expected value, we must know what the expected value is. We are fortunate in the cases of electron-electron and electron-positron scattering to have a theory, the Dirac theory of the electron, which correctly predicts these scatterings for the case of point-like interactions. The application of Dirac theory to electron-electron scattering was first worked out by C. Møller (1932), and electron-electron scattering is conventionally referred to as *Møller scattering* (Scott, 1951; Barber, 1953; Ashkin, 1954). Similarly, the application of Dirac theory to electron-positron scattering was initially worked out by H. J. Bhabha (1936), and electron-positron scattering is conventionally referred to as *Bhabha scattering* (Ashkin, 1954). The Møller scattering equation in the laboratory frame of reference, where the target particle is at rest, is (Jauch and Rohrlich, 1976, pp. 252-261)

$$d\sigma = \frac{2\pi r_O^2}{\beta^2} \cdot \frac{m_O}{T}\left\{\frac{1}{w^2}+\frac{1}{(1-w)^2}+\left(\frac{\gamma-1}{\gamma}\right)^2 - \frac{2\gamma-1}{\gamma^2}\frac{1}{w(1-w)}\right\}dw \ , \qquad (7.15)$$

where $r_O = e^2/m_Oc^2$, $\beta = v/c$, v = the velocity of the incoming electron, T = the kinetic energy of the incoming electron, $\gamma = (T+m_Oc^2)/m_Oc^2 = 1/\sqrt{1-\beta^2}$, $w = \Delta E/T$, and ΔE = the energy lost by the incident electron in the scattering process. The electron scattering angle θ is related to the fractional-energy-loss parameter w by the equation

$$\frac{2-(\gamma+3)\sin^2\theta}{2+(\gamma-1)\sin^2\theta} = 1-2w \ . \qquad (7.16)$$

In the same notation, the corresponding Bhabha equation is (Jauch and Rohrlich, *ibid.*)

$$d\sigma = \frac{2\pi r_O^2}{\beta^2} \cdot \frac{m_O}{T}\{A+B+C\}dw \ , \qquad (7.17)$$

where

$$A = \frac{1}{w^2} - \frac{\gamma^2-1}{\gamma^2}\frac{1}{w} + \frac{1}{2}\left(\frac{\gamma-1}{\gamma}\right)^2 ,$$

$$B = -\left(\frac{\gamma-1}{\gamma+1}\right)\left[\frac{\gamma+2}{\gamma}\frac{1}{w} - 2\frac{\gamma^2-1}{\gamma^2} + w\left(\frac{\gamma-1}{\gamma}\right)^2\right] ,$$

$$C = \left(\frac{\gamma-1}{\gamma+1}\right)^2\left[\frac{1}{2}+\frac{1}{\gamma}+\frac{3}{2\gamma^2} - \left(\frac{\gamma-1}{\gamma}\right)^2 w(1-w)\right] .$$

Eqs. (7.15) and (7.17) are two of the basic equations in what has subsequently become denoted as quantum electrodynamics, QED. The quantum mechanical Møller scattering equation differs from the classical Rutherford electron-electron scattering equation mainly by the addition of a spin term and an exchange term. The exchange term is required according to the rules of quantum mechanics when the two particles in the scattering are identical. The quantum mechanical Bhabha scattering equation does not contain an exchange term, since the two particles in this case are not identical, but Bhabha scattering includes a quantum-mechanical virtual annihilation component that must be taken into account (Ashkin, 1954).

When the first modern electron-electron and electron-positron experiments were carried out, at the rather modest energies of a few MeV (Scott, 1951; Barber, 1953; Ashkin, 1954), they were found to be in accurate agreement with the Møller and Bhabha scattering equations, and they demonstrated the importance of the spin, exchange, and virtual annihilation terms in these equations. Theoretical predictions were then made as to the types of deviations from QED that might be expected to occur at higher energies (Drell, 1958; Kroll, 1966; Behrends, 1974, 1981, 1982; Eichten, 1983). However, as the experiments were pushed to higher and higher energies, no deviations were observed. Experiments at a center-of-mass energy of 29 GeV (Bender, 1984; Delfino, 1985), which is an extremely high energy from the standpoint of electron-electron interactions, are still consistent with the predictions of QED. This indicates that the interaction force in Møller and Bhabha scattering is point-like in nature down to a distance of less than 10^{-16} cm (Bender, 1984, p. 521). The Møller and Bhabha equations in fact tell us even more than this. They not only give the *shapes* of the angular distributions correctly, but they also give the *absolute magnitudes* of the cross sections correctly. Hence this point-like interaction force is *purely electromagnetic*, since Eqs. (7.15) and (7.17) include only electromagnetic effects. No other kinds of forces are operating here. When we combine these results, we have a definitive answer to question (7.12*a*). We now know empirically that the $1/r^2$ force which operates in electron-electron and electron-positron scattering is strictly electromagnetic and is maintained down to distances of less than 10^{-16} cm. *This is one of the most decisive experimental results in particle physics!* It indicates very clearly that the charge radius R_E of the charge on the electron must be even smaller than the $1/r^2$ interaction radius r of Eq. (7.14), so that

$$R_E \ll 10^{-16} cm \ . \tag{7.18}$$

It also indicates to most physicists that the *total size* of the electron must be less than 10^{-16} cm. However, the main thrust of the present book is an attempt to establish the fact that this last statement is a *non sequitur*.

Having answered question (7.12*a*), we are now prepared to deal authoritatively with question (7.13*a*). We have two possibilities here. Either the electric charge e on the electron is a classical charge distribution ρ that interacts with itself, or it is not. If the former is the case, we can accurately estimate the self-energy W_E of this charge distribution, since it depends primarily on the spatial extent of the distribution. Assume that the charge e is in the form of a distribution ρ whose spatial elements interact with one another in accordance with the laws of classical electrostatics. The self-energy W_E of this charge distribution is given by the equation (Jackson, 1962, p. 588)

$$W_E = \frac{1}{2}\int\int\frac{\rho(\vec{r}_1)\rho(\vec{r}_2)}{|\vec{r}_1-\vec{r}_2|}d\tau_1 d\tau_2 \ , \tag{7.19}$$

where the integration is over the region of space τ delimited by R_E. For spherically symmetric charge distributions, Eq. (7.19) gives

$$W_E = \mathrm{A}\frac{e^2}{R_E} \ , \tag{7.20}$$

where A is a numerical constant of order unity. In particular, if the charge is in the form of a uniform surface distribution on a sphere of radius R_E, then A = 1/2; and if the charge is in the form of a uniform volume distribution, then A = 3/5 (Rohrlich, 1965, p. 125). Let us for simplicity set A = 1. The *largest* value that W_E can have is the total energy of the electron, $W_E = mc^2$. Thus, from Eq. (7.20), the *smallest* value that R_E can have is $R_E = e^2/mc^2$, which is just the *classical electron radius* $R_O = 2.82 \times 10^{-13}$ cm that was defined in Eqs. (1.7) and (3.2). Any smaller value for R_E will result in a classical electron self-energy

W_E that exceeds the observed mass of the electron. However, according to Eq. (7.18), R_E is in fact more than a thousand times smaller than R_O. Hence the answer to question (7.13a) is clearly *no*! The electric charge e on the electron is *not* a classical charge distribution that interacts with itself. This being the case, we are left with no theoretical basis for concluding anything about the self-energy W_E, since we do not have an alternate theory that can be invoked. We could in principle select any energy value for W_E between zero and mc^2. Given this situation, we are forced to turn to phenomenology, where, fortunately, an answer awaits us.

The phenomenology we have in mind was discussed in a preliminary manner in the last chapter, and is developed in detail in Chapters 9 - 14. We simply mention here the pertinent results that have a bearing on the value of the electrostatic self-energy W_E of the charge e. In order to account quantitatively for the spin of the electron, we introduce in Chapter 10 the concept of the relativistically spinning sphere. This is a sphere of matter that spins at the relativistic limit (where the equator is moving at the velocity of light), and which accurately reproduces the angular momentum properties of the electron. Then, in order to quantitatively reproduce the magnetic moment of the electron, we place a point charge e on the surface of the sphere and allow it to move freely. It is forced to the equator by the action of the magnetic forces, where it generates the observed electron magnetic moment, as we describe in Chapter 11. This model correctly reproduces the gyromagnetic ratio of the electron, and is in fact the only phenomenological model known to the author that accomplishes this result.[3] Now, the key point here is that the equatorial charge e on the relativistically spinning sphere must necessarily have zero mass, since the equator of the spinning sphere is moving at, or just below, the velocity of light, c. A finite mass for the charge e would be greatly enhanced relativistically, and would ruin the dynamical properties of the model. This model *unequivocally* requires $W_E = 0$ for the charge e.

The conclusion that the Coulomb self-energy W_E of the electric charge must be equal to zero is not unique with the present model. As we point out at the end of Chapter 15, Fokker (1929), Wheeler and Feynman (1945, 1949), and Rohrlich (1964) have all advanced theories in which this assumption is made. From the present point of view, an electric charge e does not interact with its own electrostatic field, at least when it is perched on the equator of a spinning electron.

The assumption that $W_E = 0$ raises an immediate question with respect to QED. In QED, divergent integrals are encountered that are dealt with by invoking charge renormalization and mass renormalization. Mass renormalization involves replacing, at an appropriate point in the calculations, the mass term that appears in the QED equations, which is formally infinite, with the experimental electron mass. This mass is presumably an electromagnetic mass (since QED includes only electromagnetic effects), so that if we set $W_E = 0$, we seem to be contradicting the traditional QED interpretation. However, this is not the case. The electromagnetic mass includes both electric and magnetic components. And as Jackson, for example, points out (1962, p. 593), the *electric* self-energy shown in Eq. (7.19) has a $1/r$ singularity as r gets very small, whereas QED exhibits a singularity that is logarithmic in r. We will demonstrate in the next chapter that the *magnetic* self-energy W_H is non-zero, and that the classical expression for W_H in fact contains the same logarithmic divergence that is encountered in QED. Thus the mass divergences in QED seem to be related to magnetic rather than electric phenomena, at least as viewed from the present phenomenological standpoint.

The assumption that the electric self-energy W_E of the electron is zero raises another question as well. Namely, if the mass of the electron is not due to electrostatic self-energy, then what is it due to? As mentioned above, we will demonstrate in Chapter 8 that the magnetic self-energy W_H is non-zero. However, we will also demonstrate that the magnitude of W_H can be estimated as $W_H \sim \alpha/2\pi \cdot mc^2$, or roughly only one-thousandth of the observed electron mass. Thus the bulk of the electron mass must be attributed to *non-electromagnetic* entities. Hence, if we rule out gravitational contributions, which we do at the length scales considered here ($r \geq 10^{-17}$ cm), we are consequently required to invoke a "mechanical" electron mass that is primarily responsible for the inertial properties and the total energy of the electron.

There is another interesting ramification of these results. In Chapter 3 we mentioned the concept of *Poincaré forces* (Jackson, 1962, pp. 592-593; Oppenheimer, 1970, pp. 85-89), which were postulated for the early electron models of Abraham and Lorentz in order to hold the electron together. The idea here was that the charge elements ρ of the electric charge e repel each other, and some kind of non-electromagnetic (and therefore "mechanical") force is required to overcome this electrostatic repulsion. However, we have concluded here that this electrostatic repulsion, with its consequent electrostatic self-energy W_E, does *not* occur. Thus the Poincaré forces are no longer required, at least for electrostatic effects: *the electric charge e, whatever it is, holds itself together.* Hence we can no longer invoke the need for Poincaré forces as an argument that the electron must have a non-electromagnetic "mechanical" component. But we now need this mechanical component in order to account for the total mass of the electron. And, in fact, the mechanical mass seems to perform the function of keeping the electric charge e confined in the electron, as we discuss in Chapter 11. The Poincaré mass has disappeared as a confining entity that operates inside of the charge e, only to reappear as a confining entity that encompasses and contains the point-like charge e.

It is pertinent to recapitulate here what has been accomplished thus far in our study of the *electric* properties of the electron. Starting with the facts that the electron has a mass m and a charge e, we deduced that if the mass m is due to the electrostatic self-energy W_E of the charge e, then the charge radius must be comparable with the classical electron radius, $R_O \sim 10^{-13}$ cm. Furthermore, a non-electromagnetic Poincaré force of some kind is required to hold this self-interacting charge e together. But the Møller and Bhabha scattering experiments clearly demonstrate that the electron has a charge radius $R_E < 10^{-16}$ cm. Hence the mass m cannot be attributed to W_E, at least as calculated classically, since the classical calculation would give $W_E >> mc^2$. Thus the electric charge e does *not* interact with itself, which removes the need for invoking Poincaré forces. Unfortunately, it also removes the only theoretical basis we had for assigning a value to W_E. However, if we now introduce the dynamical properties of the electron – its spin angular momentum $J = 1/2\,\hbar$ and magnetic moment $\mu = e\hbar/2mc$ – which were briefly described in the last chapter (this is the subject of Part III), we can phenomenologically conclude that $W_E = 0$.

These dynamical properties also suggest a magnetic self-energy $W_H \sim \alpha/2\pi \cdot mc^2$, which is non-zero, but which is only a tiny fraction of the observed electron mass. And they indicate that the overall radius of the electron is comparable to the Compton radius R_C (see Eq. 6.13).

We are thus led to the assumption that the electron must contain a Compton-sized non-electromagnetic *mechanical* mass component that is mainly responsible for the inertial properties of the electron. Furthermore, since the Møller and Bhabha scattering processes show that the interaction forces are purely electromagnetic, this mechanical mass must interact extremely weakly with other particles. Hence we arrive at the concept of an electron that is composed of a non-interacting mechanical mass, $m \sim 0.5$ MeV/c^2, and an embedded non-self-interacting electric charge e. The mass m gives rise to the total energy and the spin angular momentum of the electron, and the charge e gives rise to its interactions. However, this cannot be the complete story, as we can tell by extending this analysis to include the μ meson, or muon.

The electron and muon, when taken together, constitute one of the real enigmas of twentieth century particle physics.[4] These are the two lightest charged particles known, and they are clearly related to one another. In particular, the muon decay process is

$$\mu^- \rightarrow e^- + \nu + \bar{\nu} , \tag{7.21}$$

where μ, e, ν and $\bar{\nu}$ denote muon, electron, neutrino, and antineutrino, respectively. Thus the muon seems to be in some sense an excited state of the electron. Moreover, the relationship between the electron and muon goes much deeper than just the decay process shown in (7.21). The muon has a mass that is 207 times the electron mass, a spin that is equal to the electron spin, and a magnetic moment that is essentially the electron magnetic moment scaled by the inverse ratio of the masses. The scaling laws for the spins and magnetic moments of the electron and muon are properties that we in fact expect to find if we are guided by the systematics of the spinning-sphere model of Part III. But the crucial point here is that QED, which reproduces the magnetic moment of the electron to fantastic accuracy (see Chapter 9), also reproduces the magnetic moment of the muon to this same incredible level of accuracy. Both of these particles have almost the same QED Feynman diagrams in the magnetic moment calculation, and the few diagrams that are different are in fact required in order to account for slight differences in the anomalous magnetic moments of these particles (Lautrup, 1972, pp. 224-238). Thus both particles fall within the scope of QED, which clearly establishes their relationship to one another. And yet QED has no explanation for the electron-muon mass ratio, nor indeed for the fact that the muon even exists. Hence the electron-muon enigma mentioned above.[4]

Our interest here in the muon arises from the properties that we ascribe to its *mass*. As can be seen in Eq. (7.21), the electric charge e of the muon is transferred to the electron in the decay process. In view of the similarities between these two particles cited above, it seems fair to conclude that the charge e plays the same role in the muon as it does in the electron. Thus the muon must have a mechanical muon mass that is 207 times as massive as the mechanical electron mass, and which is responsible for the inertial properties of the muon. Furthermore, since QED accurately accounts for the production process

$$e^+e^- \rightarrow \mu^+\mu^- \tag{7.22}$$

as a purely electromagnetic phenomenon, the mechanical muon mass must be as non-interacting as is the mechanical electron mass. It provides the total energy of the muon (apart from small magnetic effects), and also the spin angular momentum of the muon, but it does not contribute appreciably to the interactions of the muon. However, the muon mass serves one other essential function. *It contains the "muonness" of the muon.* That is, the distinguishing feature between electrons and muons resides in the mass, and not in the charge. Specifically, the mechanical mass of the muon or electron carries not only the inertial properties of the particle, but also its *lepton number*. Muons and electrons belong to different lepton families. The *muon* and the *muon neutrino* carry the lepton number of *muons*, and the *electron* and the *electron neutrino* carry the lepton number of *electrons*. This lepton number is conserved within each family. Thus, for example, in the decay process shown in Eq. (7.21), the neutrino ν in the final state is a muon neutrino that carries the muon lepton number $+1_\mu$ which was carried in the initial state by the muon; and the antineutrino $\bar{\nu}$ in the final state is an electron antineutrino that carries the negative electron lepton number -1_E which cancels out the positive lepton number $+1_E$ carried by the final-state electron. Thus the total electron and muon lepton numbers are 0_E and $+1_\mu$ respectively, on both sides of the equation. Experimentally, the conservation of lepton number is manifested by the fact that a beam of muon neutrinos incident on a target produces only muons, whereas a beam of electron neutrinos produces only electrons. If we turn this discussion back to the electron, we see that the electron-muon dichotomy furnishes additional evidence for the conclusion, already reached above, that *the electron is more than just the charge e.*

We mention one further ramification of this discussion. As pictured above, the electron consists of a non-interacting spinning mechanical mass m and an equatorial point charge e. The mass m provides the spin angular momentum $J = 1/2\,\hbar$, and it carries the lepton number of the electron. The electric charge e generates the magnetic moment $\mu = e\hbar/2mc$ of the electron, and it provides the interaction force between the electron and other charged particles. The muon is similarly constructed. From this viewpoint, if we removed the charge e from an electron or a muon, we would be left with a particle that has the following properties: *(1)* zero charge, *(2)* zero magnetic moment, *(3)* spin $J = 1/2\,\hbar$, *(4)* lepton number equal to that of the electron or muon, and *(5)* a vanishingly small interaction cross section. These are the properties that are ascribed to *neutrinos*. The one major problem with this conceptual procedure of removing the charge e and thus changing an electron or a muon into a neutrino arises with respect to the mass values. The masses of the neutrinos have not been determined, and are presently the subject of intense experimental investigation. It is known that the mass of the electron neutrino, if it is non-zero, is vastly smaller than the mass of the electron; and the mass of the muon neutrino, if it is non-zero, is vastly smaller than the mass of the muon. Thus the charge on the electron is in some manner related to the stability of the mechanical mass of the electron, and similarly for the muon. Sidestepping the problem of the neutrino mass values, the main conclusion we would like to draw from this discussion is that the concept of an essentially *non-interacting* mechanical particle mass, to which we are led on the basis of the present studies, is one which is in fact within the domain of existing elementary particle phenomena.

At the beginning of this chapter, we distinguished between the intrinsic size R_E of the charge e on the electron, and the charge distribution size R_{CD} that is manifested experimentally. The only types of experiments we have discussed thus far are the Møller and Bhabha particle scattering experiments, which yield an extremely small value for R_{CD} ($< 10^{-16}$ cm), and hence also for R_E (since $R_E \leq R_{CD}$). But there is another type of phenomenon, the Lamb shift, that yields a much larger value for R_{CD}, a value roughly a million times larger. In the Dirac theory of the hydrogen atom, the $S_{1/2}$ and $P_{1/2}$ levels should have exactly the same energy. However, spectroscopic results in the 1930s revealed an anomaly in hydrogen,[5] which Pasternack (1938) identified as a possible upward energy shift of the $2S_{1/2}$ level relative to the $2P_{1/2}$ level. In the late 1940s, Lamb and coworkers (Lamb, 1947), using newly developed microwave techniques, succeeded in obtaining accurate measurements of the splitting of these levels. Theoretical estimates of this splitting were obtained by Bethe (1947) and Welton (1948), and were soon followed by full-blown QED calculations.[6] The physical idea behind the Lamb shift is that the electron continually interacts with the background radiation field and is buffeted back and forth in space. It exhibits *zitterbewegung* motion. Thus it does not appear as a point charge, but rather as a charge that is smeared out in space over a distance (Dyson, 1951) that is comparable to the Compton wavelength of the electron, $R_C \sim 4 \times 10^{-11}$ cm.[7] The S-state electrons, which pass very close to the atomic nucleus, have atomic Coulomb forces that are decreased due to this smearing out of the charge, so these electrons have weaker binding energies. The P-state electrons, on the other hand, have wave functions that vanish at the origin. These electrons never approach the nucleus, and their Coulomb potentials (which we attribute here to the *quantum current loops* of Eq. 7.10) are essentially unaffected by the radiative recoils. Hence the S states are shifted upward in energy relative to the P states, in agreement with Pasternack's hypothesis.

The electron bound-state charge distribution radius $R_{CD} \sim R_C$ that is deduced from the Lamb experiments is very large. It is the radius that we have denoted as R_{QED} in Eqs. (1.1e) and (1.6). The magnitude of the Lamb shift is accurately reproduced by the equations of quantum electrodynamics.[6] However, the physical interpretation of these QED equations is somewhat ambiguous. As Milonni, for example, points out (Milonni, 1980, pp. 1-21), Bethe's calculation of the Lamb shift (Bethe, 1947) indicates that interactions of the electron with the Coulomb field of the proton are responsible for the effect, whereas Welton's calculation (Welton, 1948) suggests that interactions of the electron with the background radiation of the spatial vacuum are responsible. Quantum-mechanically, the differences in these two viewpoints can be traced to the ordering that is used for the atomic and vacuum-state field operators. Mathematically, these field operators commute, so that the ordering does not affect the results of the calculations. But it does affect the physical interpretation that is ascribed to the equations. The interesting fact for our present purposes is that in the hydrogen atom, the charge on the electron appears to be spread out over a very large region of space as compared to the intrinsic size R_E of the charge itself. The

QED calculations accurately give the magnitude of the effect, but QED itself, as we have just seen, does not offer a clear-cut explanation for the physical process behind the effect. In the electron model that we have been led to – a point charge embedded in a spherical mechanical mass – the radius of the mechanical sphere is comparable to the Compton radius R_C, as we will demonstrate in detail in Part III. Thus the large size that is deduced on the basis of the Lamb shift for the spatial distribution of the electron charge may be due, at least in part, to the fact that the electron actually *is* that large. But there is another factor operating here which may serve to disguise this large electron size, as we now discuss.

There is a crucial question that has been hanging over all of the results developed thus far. *If the electron actually is a very large Compton-sized object, how can it exhibit point-like scattering?* The interactions of this large electron arise from a point charge e that is embedded on the equator of a non-interacting mechanical mass m. Thus, even if the charge e is point-like, as it must be, and even if it is entirely responsible for the interaction forces, as experiments indicate, it doesn't seem that this electron model should lead to point-like scattering. But it does! And the reason can be found in the *Golden Rule of Rigid Body Scattering*, which is described in Chapter 15. The idea behind this Golden Rule is the following. When the calculation of the dynamics of the relativistically spinning sphere is carried out, which we do in Chapter 10, it is found that phenomenologically correct answers are obtained by completely neglecting internal strains in the rapidly spinning sphere. But the only way we can have zero internal strains is in the case of a "rigid body", which transmits applied stresses essentially instantaneously to all parts of the body. Thus the mechanical mass that we have invoked for the electron has an important postulated property: *the mechanical electron mass has the attributes of a rigid body.* We now apply the Golden Rule of Rigid Body Scattering, which is in fact a standard result contained in books on mechanics, and which states that any external force applied to a rigid body can be decomposed into a translational component that acts on the *center of mass* of the body, plus a torque that acts as a couple around the center of mass. When this result is applied to electron scattering, it leads in general to point-like angular distributions, as we describe in Chapter 16. But there is a narrow energy window, in the keV region, where finite-size effects can appear. Since this keV energy window occurs in an energy domain that has not been thoroughly explored experimentally, and since there has been no previous theoretical motivation to investigate this keV region in detail, the experimental discrepancies which do exist in this region (see Chapter 17) have been generally overlooked. The Golden Rule of Rigid Body Scattering may also be operative in atomic orbitals, where it may serve to cancel out some finite-electron-size contributions to the Lamb shift.

As a final comment in this chapter, we should point out that the *zitterbewegung* motion revealed by the Lamb shift is not the only phenomenon that indicates a large electron charge distribution radius, $R_{CD} = R_{QED}$. *Vacuum polarization* is another standard QED effect, and it leads to a Coulomb polarization of the vacuum state by the charge e, where this polarization extends over a distance that is comparable to the Compton radius R_C (Blokhintsev, 1973, pp. 95-99). Thus, whereas the actual radius R_E of the electric charge e

is point-like, the distribution radius R_{QED} of this charge appears, in at least some experimental situations, to be vastly larger. Vacuum polarization contributions to the Lamb shift are of opposite sign to the zitterbewegung contributions, and are much smaller.[6]

By studying just the *electric* properties of the electron, and by taking the implications of these properties seriously, we have obtained a great deal of information about the possible structure of the electron. In the next chapter we turn to the *magnetic* properties of the electron. These properties, which tell a story in their own right, serve to reinforce the picture that we have created here. Also, they seem to shed some light on the physical phenomena that form the underlying structure of quantum electrodynamics.

Notes

1 This result was first made known to the present author when it was assigned as a test problem by Prof. G. Uhlenbeck in a graduate course on electricity and magnetism given at the University of Michigan in 1950.

2 This suggests that the spin quantization angle θ_{QM} has a dynamical basis (see Chapter 13).

3 See Note 4 in Chapter 2.

4 The τ meson also belongs to this family of leptons. The same general remarks that apply to the electron-muon dichotomy can also be extended to encompass the τ.

5 For a good review of the early history of the Lamb shift, see Lamb (1951).

6 For a summary of Lamb shift QED results, see Lautrup (1972, pp. 198-209).

7 The extent of the spatial recoil of an electron due to the emission and reabsorption of virtual photons can be estimated, using the Heisenberg Uncertainty Principle to relate the photon energy to its virtual lifetime. The calculated electron spatial recoil distances are comparable to the Compton radius R_C.

Chapter 8

The Magnetic Size, Magnetic Self-Energy and Anomalous Magnetic Moment of the Electron

The electron was discovered around the turn of the century, and for a quarter of a century only its static properties – its mass m and charge e – were known. Using just these properties, physicists were led to the conclusion that the natural size for the electron is the classical electron radius, $R_O = e^2/mc^2 = 2.82 \times 10^{-13}$ cm, as we discussed in Chapter 3. Then the dynamical properties of the electron – its spin J and magnetic moment μ – emerged (Uhlenbeck and Goudsmit, 1926). The manner in which this occurred historically was described in Chapter 3, and some of its classical implications were considered in Chapter 6. Fermi was perhaps the first to recognize that these dynamical properties have profound ramifications with respect to the size of the electron. Working together with Rasetti (1926), he demonstrated[1] that the magnetic self-energy W_H associated with the magnetic moment μ sets a lower bound on the size of the electron, and this lower bound is considerably larger than R_O. The suggestion was then made by these workers that[1]

> "a possible solution to such a paradox was to consider the electron magnetic structures as much greater than the electric-charge distribution".

One electromagnetic configuration which is in obvious agreement with this suggestion is a current loop formed by a rotating point-like charge.

It is of interest here to reproduce Rasetti and Fermi's calculation of the magnetic self-energy W_H of the electron (Rasetti and Fermi, 1926). This is the energy that is contained in the magnetic field associated with the magnetic moment of the electron. Since this magnetic field, according to Ampere's hypothesis (Feynman, 1961a, p. 84), arises from the *motion* of the electric charge e, and is asymptotically independent of the point-like charge radius R_E (Eq. 7.18), we expect the *magnetic* self-energy W_H to be more amenable to a classical treatment than is the *electric* self-energy W_E (Eq. 7.19), which we concluded in Chapter 7 must be equal to zero. Let us represent the magnetic moment of the electron by a current loop. Using polar coordinates r, θ and z, and orienting the axis of the current loop along the z axis, we have the asymptotic magnetic field components (Jackson, 1962, p. 143)

$$H_r = \frac{2\mu\cos\theta}{r^3}, \qquad H_\theta = \frac{\mu\sin\theta}{r^3}, \tag{8.1}$$

where μ is the magnetic moment of the electron. As can be seen, these field components do not depend on R_E. Assuming that these asymptotic fields apply all the way in to a *magnetic* radius R_H, we obtain an *external* magnetic self-energy

$$W_H^{EXT} = \left(\frac{\mu^2}{8\pi}\right)\int_{R_H}^{\infty}\int_0^{\pi}\left(\frac{1}{r^6}\right)(3\cos^2\theta + 1)\,2\pi r^2 \sin\theta\, d\theta\, dr = \frac{\mu^2}{3R_H^3}, \quad (r > R_H)\ . \tag{8.2}$$

This calculation assumes that the magnetic moment of the electron is spread throughout the spherical volume defined by R_H. Several years later, and quite independently, Born and Schrödinger (Born, 1935) made a similar calculation of W_H, but with the assumption that the electron's magnetic moment is distributed on the surface of the sphere. This calculation gave

$$W_H^{EXT} = \frac{\mu^2}{2R_H^3}, \tag{8.3}$$

in close agreement with the Rasetti and Fermi result.

We can extend the Rasetti and Fermi calculation of W_H by making an estimate of the energy that is contained *inside* the magnetic radius R_H. To accomplish this, we use Eq. (8.1) to calculate the fields that are present at R_H, and we then assume that these magnetic field values remain constant (independent of r) throughout the interior of the sphere. Since the magnetic fields should actually increase at small distances, this procedure represents a plausible lower limit for the magnetic self-energy in this region. Carrying out an integration similar to that of Eq. (8.2), we obtain the following lower bound for the *internal* magnetic self-energy:

$$W_H^{INT} \geq \frac{\mu^2}{8\pi R_H^6}\int_0^{R_H}\int_0^{\pi}(3\cos^2\theta + 1)\,2\pi r^2 \sin\theta\, d\theta\, dr = \frac{\mu^2}{3R_H^3}, \quad (r < R_H)\ . \tag{8.4}$$

Adding Eqs. (8.2) and (8.4) gives

$$W_H^{TOT} \geq \frac{2\mu^2}{3R_H^3} \tag{8.5}$$

as a lower bound for the *total* magnetic self-energy of the electron. To see what value this implies for R_H, we now set the magnetic energy equal to the total electron energy, so that $W_H = mc^2$, and we insert the zeroth-order value for the electron magnetic moment,

$$\mu = \frac{e\hbar}{2mc}\ . \tag{8.6}$$

With these substitutions, Eq. (8.5) becomes

$$R_H^3 \geq \left(\frac{\alpha}{6}\right) R_C^3\ , \tag{8.7}$$

where $\alpha = e^2/\hbar c$ is the fine structure constant, and $R_C = \hbar/mc$ is the electron Compton radius. Numerically, we have

$$R_H \geq 0.106\,R_C = 4.09 \times 10^{-12}\,\text{cm}\ . \tag{8.8}$$

This lower limit for the magnetic radius R_H that we thus obtain from the extended Rasetti and Fermi calculation is a little over ten times the classical electron radius R_O, so that the "classical electron" is indeed much larger *magnetically* than it is *electrically*, as Fermi and Rasetti had noted (Rasetti, 1926). Also, this lower bound for R_H is almost exactly one-tenth the Compton radius R_C. As we will see shortly, this is a significant numerical result!

Eq. (8.6) is the equation for the magnetic moment of the electron that was postulated by Uhlenbeck and Goudsmit (1926). As we discussed in Chapter 3, this equation was initially deduced on an *ad hoc* basis. It required the Thomas precession (Jackson, 1962, pp. 364-365) to make the formalism consistent, and the Dirac equation to give it a theoretical rationale. Eq. (8.6) is in fact a very remarkable equation: the only parameter it contains which is specific to the electron is the mass m. Now, the magnetic moment is an electromagnetic phenomenon, and its value, at least from a classical viewpoint, involves just the charge e, a current loop radius r, and an angular velocity ω (see Eqs. 6.7 - 6.9). Hence the appearance of the mass m in Eq. (8.6) suggests that it is merely serving as a parameter which is in some manner related to the electromagnetic quantities e, r, and ω. (The nature of this relationship will become clear in Chapters 10 - 14.) The essential point here is that the magnetic moment of the electron is in fact accurately characterized by a single parameter – its mass m. This same relationship also holds for the muon, so that we have

$$\frac{\mu_\mu}{\mu_e} = \frac{m_e}{m_\mu} \tag{8.9}$$

as an extremely accurate equation.[2] This is a scaling by more than two orders of magnitude in the mass m. In fact, as we discussed in Chapter 5, this scaling can be further extended to encompass the "intrinsic magnetic moments" (per unit charge) of the nucleon constituent quarks u and d (Feld, 1969, p. 339) and the charmed quark c (Gottfried and Weisskopf, 1984, p. 112). This then becomes a scaling by a factor of 3000 in the masses and magnetic moments of these supposedly unrelated types of "particle" – the electron and muon on one hand, and the hadronic quarks on the other. Hence Eq. (8.6) expresses a very general relationship between masses and magnetic moments.

About two decades after Eq. (8.6) for the electron was set forth by Uhlenbeck and Goudsmit (1926), physicists discovered that it was not completely correct. The actual magnetic moment, for both electrons and muons, is approximately 0.1% larger than this value (Lautrup, 1972, pp. 216-217):

$$\mu \cong \frac{e\hbar}{2mc}\left(1+\frac{\alpha}{2\pi}\right), \tag{8.10}$$

where $\alpha \sim 1/137$ is the fine structure constant, and $\alpha/2\pi$ is the famous Schwinger correction term (Schwinger, 1948). This raises an immediate question. As Eq. (8.6) shows, the magnetic moment of a lepton depends on a single parameter – the mass m. And as Eq. (8.9) shows, the magnetic moments of different leptons scale with this same mass parameter. Hence we logically have the following question: *is the anomalous magnetic moment of the electron related to a mass component in the electron?* That is, should we write Eq. (8.10) in the form

$$\mu \;=\; \frac{e\hbar}{(m-\Delta m)c}\,, \tag{8.11}$$

where

$$\Delta m \;\cong\; m \cdot \frac{\alpha}{2\pi}\;? \tag{8.12}$$

The evidence for this hypothesis is certainly suggestive. And, interestingly, we can pinpoint exactly what the mass component Δm must be. The reason we can do this follows from the fact that this anomalous component occurs not only in the magnetic moment μ of the electron (Eq. 8.10), but also in the gyromagnetic ratio g of the electron (see Eq. 3.5),

$$g \;=\; \left(\frac{2mc}{e}\right)\left(\frac{\mu}{J}\right), \tag{8.13}$$

where $J = 1/2\,\hbar$ is the spin angular momentum of the electron. Inserting the values for μ and J into Eq. (8.13), we obtain

$$g-2 \;\cong\; 2 \cdot \frac{\alpha}{2\pi}\,. \tag{8.14}$$

The (g-2) ratio has been measured experimentally to a high degree of accuracy for electrons (Wesley, 1971) and for muons (Bailey, 1979). The agreement of these measurements with their corresponding theoretical calculations constitutes one of the most impressive successes of QED (see Chapter 9). The significance of these results with respect to the present discussion is the following: since the gyromagnetic ratio g exhibits the same anomaly as does the magnetic moment, it means that this anomaly is occurring in the magnetic moment μ, but *not* in the spin angular momentum J. This fact is brought out clearly in Eq. (8.13). Hence if this anomaly is related to a mass component in the electron, it must be a mass component that does *not* contribute to the spin angular momentum J. Now, there are four different types of mass (or, equivalently, energy) that we can logically attribute to the electron:

(a) electrostatic self-energy W_E ;
(b) magnetic self-energy W_H ;
(c) mechanical mass W_M ;
(d) gravitational mass W_G .

Of these four possible mass components, only W_H is irrotational, and hence does not contribute to J. The mass W_M certainly contributes to J. The masses W_E and W_G can in principle also contribute. However, we decided in Chapter 7 that we must have $W_E = 0$. And W_G is totally negligible at the length scales considered here ($R > 10^{-17}$ cm).[3] Thus W_H is singled out uniquely as the Δm mass component in Eq. (8.12), so that we must have

$$W_H \;=\; c^2\Delta m \;\cong\; mc^2 \cdot \frac{\alpha}{2\pi} \;=\; 593 \text{ eV}\,. \tag{8.15}$$

But is 593 eV a plausible value for the magnetic self-energy of the electron? We obtain an immediate affirmative answer to this question by noting that the factor of $\alpha/2\pi$ which occurs in Eq. (8.15) is essentially identical to the factor $\alpha/6$ which occurs in Eq. (8.7). This tells us that if we set the magnetic radius R_H in Eq. (8.5) equal to the Compton radius

R_C, which is our intuitively expected size for the electron (Chapters 5 and 6), then the calculated value for the magnetic self-energy W_H is

$$W_H \ \geq \ \frac{\alpha}{6} \cdot mc^2 \ = \ 621 \text{ eV}, \quad (R_H = R_C), \tag{8.16}$$

which is almost identical to the mass value of 593 eV shown in Eq. (8.15). Hence the magnetic self-energy that we calculate for a Compton-sized electron by using the estimate of Rasetti and Fermi (1926) is of just the right magnitude to account for the observed anomaly in the magnetic moment of the electron. Of course, the Rasetti and Fermi calculation is only a lower bound for W_H. Furthermore, the use of the Compton radius for the electron is not quite correct phenomenologically, as we discuss later on in the chapter. (These two *caveats* tend to offset one another.) But the key point here is the following: *the estimated value for the magnetic self-energy of the electron is about 0.1% of the total energy of the electron; and the anomaly in the magnetic moment of the electron is a 0.1% anomaly!*

We have established the following sequence of results: ***(a)*** The equation for the magnetic moment of the electron, Eq. (8.6), contains only one electron-specific parameter, the electron mass m. ***(b)*** The scaling of lepton (and quark) magnetic moments is in terms of this same mass parameter (Eq. 8.9). ***(c)*** Guided by *(a)* and *(b)*, it seems logical to relate the observed $\alpha/2\pi$ anomaly in the magnetic moment of the electron (Eq. 8.10) to an anomalous mass component $\Delta m = m\alpha/2\pi$, as shown in Eq. (8.11). ***(d)*** Since this same $\alpha/2\pi$ anomaly occurs in the gyromagnetic ratio g of the electron (Eq. 8.14), the electron spin J does *not* contain the anomaly, which indicates that the anomalous mass component Δm is irrotational in nature. ***(e)*** Of the possible types of electron mass or energy that we can envision, only the magnetic self-energy is irrotational, which essentially mandates that Δm must correspond to the magnetic self-energy of the electron. ***(f)*** Arguments we have already presented (Chapters 5 and 6) and arguments we will present later (Chapters 10 - 14) suggest that the electron is a Compton-sized particle. ***(g)*** If we take the extended Rasetti and Fermi estimate (Eq. 8.5) for the magnetic self-energy W_H of the electron and extrapolate it to apply to a Compton-sized electron, we obtain a value for W_H (Eq. 8.16) which accurately corresponds to the anomalous Δm mass term (Eq. 8.15) that we deduced on the basis of results *(a)* - *(c)*. Thus the picture we obtain from this chain of reasoning is self-consistent. And the significant conclusion it tells us is that *the anomaly in the magnetic moment of the electron arises as a consequence of the magnetic self-energy of the electron* (in a manner that will be described in detail in Chapter 14). This conclusion appears to be original with the present author (Mac Gregor, 1987 and 1989).

As we have already mentioned several times, there exists an excellent theory, the formalism of quantum electrodynamics, which accurately reproduces the anomalous magnetic moment of the electron. In fact, this QED calculation constitutes one of the most impressive achievements that has ever been obtained in all of science (see Chapter 9). The results that we have described in the present chapter, on the other hand, while quantitative in nature, can hardly claim a QED-like level of accuracy. This raises two questions, one of them technical, and the other more pragmatic. The technical question is the following:

how does the picture that is presented here fit in with the formalism of QED? The pragmatic question, which also has a bearing on the technical question, is one of justification: *if QED works so well, why do we even have to think about a visual model of the electron – who needs it?* This pragmatic question was in fact taken up by Feynman, long before the present work, and he gave a very convincing answer. In a summary talk on the status of QED, he remarked as follows (Feynman, 1961a, pp. 75-76):[4]

> "It seems that very little physical intuition has yet been developed in this subject. In nearly every case we are reduced to computing exactly the coefficient of some specific term. We have no way to get a general idea of the result to be expected. To make my view clearer, consider, for example, the anomalous electron moment ... . We have no physical picture by which we can easily see that the correction is roughly $\alpha/2\pi$; in fact, we do not even know why the sign is positive (other than by computing it). ... We have been computing terms like a blind man exploring a new room, but soon we must develop some concept of this room as a whole, and to have some general idea of what is contained in it. As a specific challenge, is there any method of computing the anomalous moment of the electron which, on first rough approximation, gives a fair approximation to the α term ... ?"

The challenge set forth here by Feynman is to devise some kind of an electron model which accounts for both the sign and the magnitude of the anomalous magnetic moment, and hence provides a physical underpinning for the Feynman diagrams that are used in the QED calculation of this quantity. The model we have sketched here is a partial answer to Feynman's challenge. It gives, via Eqs. (8.11), (8.15), and (8.16), an estimate for the *magnitude* of the anomalous magnetic moment of the electron. However, we have not yet accounted for the *sign* of this effect. In order to accomplish this, we need the systematics that is developed in Chapters 10 - 14. But we can already see the usefulness of having some kind of a classical understanding as to what the anomalous magnetic moment of the electron is all about.

As a final topic in this chapter, we must deal with the technical question raised above. That is, we must attempt to relate the present semi-classical calculations to the mathematical formalism of quantum electrodynamics. Since QED encompasses phenomena such as vacuum polarization which are beyond the scope of the present studies, we cannot pursue this topic in detail. But we can at least make a few observations. There is one magnetic calculation we have not yet carried out, and which is central to this discussion. This is the calculation of the magnetic self-energy of a current loop. The estimate of the magnetic self-energy W_H of the electron that was made by Rasetti and Fermi (1926) was obtained by taking the known asymptotic fields of a magnetic dipole (Eq. 8.1) and integrating these asymptotic forms all the way in to an assumed magnetic dipole radius R_H. Having no model for the electron, and being unaware of the (as yet undiscovered) anomaly in the magnetic moment, they had no basis for pinning down the value of R_H, other than to say it must be large enough so that the magnetic self-energy W_H does not exceed the total energy mc^2 of the electron. This line of reasoning leads to a lower limit for R_H (Eq. 8.8) that is just about 1/10 the electron Compton radius. Guided by our subsequent knowledge of the magnetic moment anomaly, and also by the arguments developed in Chapters 5 and 6 that favor a Compton-sized electron, we proceeded to set $R_H = R_C$, which yielded a Fermi-type estimate for W_H (Eq. 8.16) that matches the magnitude of the observed magnetic anomaly (Eq. 8.15). If we now assume that the magnetic moment of the electron arises from a Compton-sized equatorial current loop, which is what we expect from Ampere's hypothesis, we can extend the above discussion by making a direct calculation of the magnetic self-energy that is associated with this current loop. This calculation displays some intriguing QED-like characteristics. Let us represent the current loop as a thin wire of radius R_E which is bent into a circle of radius R_C, where $R_C >> R_E$. The self-inductance L of this current loop is (Smythe, 1939, p. 316, Eq. 2)

$$L = 4\pi R_C \left\{ \eta \left(\ln 8 \frac{R_C}{R_E} - 2 \right) + \frac{1}{4} \eta' \right\} \equiv 4\pi R_C \cdot B \ , \tag{8.17}$$

where η and η' are the permeabilities outside of and inside of the thin wire, and where B denotes the terms in the brackets. The self-energy W_H of this current loop is

$$W_H = \frac{1}{2} L i^2 = \frac{1}{2} L \left(\frac{e}{2\pi R_C} \right)^2 , \tag{8.18}$$

where

$$i = \frac{e}{2\pi R_C} \tag{8.19}$$

is the electric current, and where the charge e is assumed to be moving at the velocity of light c (see Eq. 6.8). If we now insert the value for L from Eq. (8.17) into Eq. (8.18), we obtain

$$W_H = \frac{e^2}{2\pi R_C} \cdot B = \frac{\alpha}{2\pi} \cdot mc^2 \cdot B \ . \tag{8.20}$$

Thus if we set the bracket term B equal to unity, we obtain the Schwinger magnetic moment correction factor $\alpha/2\pi$ as a direct result of this calculation! Unfortunately, we have as yet no theoretical basis for setting $B = 1$, but it is nevertheless intriguing to see this very complicated QED result emerge from such a simple calculation, especially since this calculation is based on an essentially classical representation of the physical phenomenon that is involved.

If this were all there was to the story, it might be dismissed as just an interesting coincidence. But there is more. Let us now examine the bracket B. The factor R_E that appears in the denominator of the logarithm in Eq. (8.17) is the radius of the electric charge e, which we know experimentally to have the value $R_E < 10^{-16}$ cm (Eq. 7.18). Thus it is much smaller than the factor $R_C = 4 \times 10^{-11}$ cm (Eq. 1.1) that appears in the numerator of the logarithm. Hence this divergent logarithmic term is dominant, and we see from Eqs. (8.17) and (8.20) that

$$W_H \sim \alpha \cdot mc^2 \cdot \ln \frac{R_C}{R_E} \ . \tag{8.21}$$

This equation is of precisely the same form as the logarithmic singularity that appears in the QED calculation of the electromagnetic self-energy of a particle, as is illustrated for example in Jackson's book (1962, p. 593). This tells us two closely related facts about QED: *(1)* the logarithmic QED divergence that occurs in the electromagnetic self-energy of a particle is a *magnetic* divergence, and not the *electrostatic* divergence that is commonly assumed; *(2)* this logarithmic divergence indeed arises as a consequence of the small spatial size of the electric charge e, but this divergence is manifested by its effect on the magnetic field of the electron rather than as an electrostatic self-interaction. In detail, the point-like rotating charge e leads to an extremely large magnetic field in its vicinity, with a consequently large value for the magnetic self-energy W_H, but the charge e does not interact with itself ($W_E = 0$).

There is one final aspect to this story about the magnetic self-energy of the current loop on the electron. Since we know the value for W_H, as deduced from the magnetic moment anomaly, we can work backwards and see what this implies for the quantities that are contained in the bracket B in Eq. (8.17). In particular, we want to see what we can find out about the radius R_E, which is the one truly unknown quantity that is involved. We start by setting $B = 1$, since this gives the correct value for the magnetic moment anomaly (Eq. 8.20). Then we set $\eta = 1$ (the permeability of free space), and we set $\eta' = 0$ (since there is in fact no wire). With these assumptions, the expression for B reduces to $\ln (8\, R_C / R_E) = 3$, which yields the following "effective charge radius":

$$R_E^{\text{eff}} \sim \frac{2}{5} R_C \quad = \quad 1.5 \times 10^{-11} \text{ cm}. \tag{8.22}$$

We thus have a paradox. Starting with a point-like value for R_E (10^{-16} cm or less), we used the assumption that $R_C >> R_E$ to derive Eq. (8.17). Applying this equation, we obtained an expression for the self-energy W_H of the current loop (Eq. 8.20). Then, by matching W_H to the magnetic moment anomaly (Eq. 8.15), we found that a much larger "effective value" for R_E (10^{-11} cm) is actually required in Eq. (8.17). But this is in fact what we should expect from QED. The factor that seems to be operating here is the QED phenomenon of *vacuum polarization*. When a point-like electric charge is placed at a certain position in space, it polarizes the surrounding vacuum state, thereby masking some of the effect of the charge. Studies of this vacuum polarization effect (Blokhintsev, 1973) indicate that it extends over a distance which is comparable to the Compton radius R_C. This is just the result that we see in Eq. (8.22), where the effective value for R_E exhibits a Compton-like size. Thus several facets of QED have emerged from this semi-classical current loop calculation: the Schwinger $\alpha/2\pi$ correction term; the characteristic QED logarithmic divergence; and a manifestation of the effect of vacuum polarization. In terms of the production of magnetic fields, vacuum polarization may be operating here in the form of a magnetic saturation effect that occurs in the immediate vicinity of the electric charge e.

We are not yet finished with this electron magnetic self-energy analysis. A minor contradiction in these results is waiting to be accounted-for. We saw in Eq. (8.15) that a magnetic self-energy of 593 eV is required in order to reproduce the magnetic moment anomaly. And we saw in Eq. (8.16) that a Compton-sized current loop gives a Fermi estimate of 621 eV for the magnetic self-energy. But this extended Fermi estimate, which is based on Eq. (8.5), is necessarily a lower limit, since it assumes that the magnetic field intensity is a constant (independent of r) inside of the radius R_C. Thus the value shown in Eq. (8.16) should be *smaller* than the value shown in Eq. (8.15). In order to account for this discrepancy, we must introduce one aspect of the problem that we have heretofore been neglecting. As we discuss in Chapters 13 and 14, the *total* quantum mechanical electron magnetic moment μ is actually a factor of $\sqrt{3}$ larger than the value used here, which is in fact the *projection* μ_Z of μ along the z axis of quantization (and which is the observable quantity). Correspondingly, the actual electron radius R is $\sqrt{3}$ larger than R_C. If we insert these larger values for μ and R into Eq. (8.5), then the calculated W_H value of 621 eV shown in Eq. (8.16) changes into

$$W_H \geq 359 \text{ eV}, \quad (R_H = \sqrt{3}R_C) . \tag{8.16'}$$

This value is now smaller than the value of 593 eV shown in Eq. (8.15), as it should be. Furthermore, in the current-loop calculation of W_H, we must now set B = $\sqrt{3}$ in order to match the $\alpha/2\pi$ magnetic anomaly. With this change, the calculated value of the effective charge radius R_E becomes

$$R_E^{\text{eff}} = 1.3 \times 10^{-11} \text{ cm} . \tag{8.22'}$$

Since the radius of the current loop is now $\sqrt{3}\ R_C = 6.7 \times 10^{-11}$ cm, which is approximately five times as large as the effective value of R_E shown here, Eq. (8.17) should still be valid.

There is still another conclusion to be drawn from these calculations. The value of 359 eV that we obtain for the magnetic self-energy W_H of the electron from the Fermi estimate is smaller than the value of 593 eV that is required empirically to match the magnetic moment anomaly, but it is not very much smaller. This indicates that the magnetic fields in the interior of the electron are in fact fairly uniform. The violent singularity that we might expect to find in the vicinity of the point charge e is *not* occurring. This indicates in turn that the effective charge radius of 1.3 x 10^{-11} cm shown in Eq. (8.22′), which was obtained from the current loop calculation of Eq. (8.17), is a reasonable one. Hence the Fermi approach to W_H, which is embodied in Eq. (8.5), and the current-loop approach to W_H, which is embodied in Eq. (8.17), give mutually consistent results.

The QED calculation of the magnetic moment anomaly in the electron contains many Feynman diagrams, which combine together to represent the physical factors that occur during the interaction of the electron with a magnetic field. If the present electron model has validity, then these same factors must be occurring here. However, as we have just seen in the example of the current loop with its expanded effective charge radius, these factors can combine together in a non-linear manner. Thus the task of relating various calculational features of the present model to specific Feynman diagrams may not be a straightforward procedure.

One final comment. The second-order magnetic moment Feynman diagram (Lautrup, 1972, p. 216, Fig. 4.1), which is the one that gives rise to the Schwinger $\alpha/2\pi$ correction term, corresponds to the process in which a single virtual photon is emitted and subsequently reabsorbed while the electron is interacting with an external magnetic field. In a discussion on the interpretation of electron mass renormalization diagrams, Feynman (1961b, p. 140) points out that the single-virtual-photon diagrams are suggestive of a correction term to the mass of the electron. Thus the procedure introduced here of identifying the anomalous magnetic moment of the electron with a mass term in the electron (Eqs. 8.11 and 8.12) seems to be in accordance with the spirit of quantum electrodynamics.

We have now completed our introduction to the topic of *The Enigmatic Electron.* In Part I we briefly reviewed the manner in which the electron transcends the bounds of classical physics as it existed *circa* 1905, and we then went on to argue that some classical concepts may nevertheless still apply at the level of the electron. In fact, the labels *classical* and *quantal* must clearly apply to two limiting versions of a single all-embracing "theory of physics". We cannot draw boundaries between these two limits on the basis of size, nor on the basis of Planck's constant h. The electron seems to inhabit a no-man's land between the classical and quantum domains, and it may be simple enough that it will serve as a guide in constructing a bridge between them. In Part II we brought out the fact that the properties of the electron suggest it is in actuality a Compton-sized particle, in spite of its propensity for scattering in a point-like manner. When we examine the few elementary particles whose sizes have actually been measured, we discover that they are all Compton-sized. The point-particle is a useful construct, especially as viewed macroscopically, but it may bear no relationship to reality. Having considered all of this background information, we now move on in Parts III and IV to the construction of an electron model that seems capable of tying these various pieces of the puzzle together. What we end up with is a large relativistically spinning mechanical sphere with a point-like charge on its equator. As we demonstrate in Part III, this model accurately reproduces the spectroscopic properties of the electron. And, as we calculate in Part IV, this large electron scatters like a point at most energies. However, there may be a multi-keV "window of opportunity" in which this Compton-sized electron reveals its true size.

Notes

1 A recent discussion of this early work is given by Belloni (1981).

2 See Lautrup (1972), p. 238. The magnetic moment versus mass scaling law shown in Eq. (8.9) has an experimental accuracy of about 7 parts in a million.

3 In heterotic string theories, the electron is assigned a length scale of about 10^{-33} cm. Only in these very small spatial domains do masses comparable to those of the electron have appreciable gravitational forces.

4 In the complete quotation, Feynman would like this model to be capable of generating higher-order correction terms as well. This is a more difficult undertaking, since some of these higher-order corrections involve phenomena such as vacuum polarization which are not quantitatively taken into account in the present semi-classical approach.

Part III

The Spectroscopic Electron

"If I can't picture it, I can't understand it."

Albert Einstein[1]

Chapter 9

Do We Need a Spectroscopic Model of the Electron?

A *spectroscopic particle model* is a model that (1) corresponds to known physical laws, and (2) enables us to calculate, or at least correlate, the spectroscopic properties of the particle. Let us ask a question:

Do we need such a model for the electron?

Speaking pragmatically, the answer up to the present time is clearly *no*! The electron was isolated and identified almost a hundred years ago, and it has led to enormous scientific advances. These advances have not required any sort of model for the electron. Thus let us ask another question:

Would it be useful to have a spectroscopic electron model?

This is a subjective question, and a research scientist and a physics teacher might well give different answers: the teacher would be delighted to have a scheme for correlating the various properties of the electron, but the researcher would possibly conclude that such a model is an unnecessary piece of baggage (see the quotation by Margenau at the end of Chapter 3). Hence let us ask still another question:

What can we calculate with a spectroscopic electron model?

Now we are seemingly on more solid ground. Either we can calculate something, or we can't. If the calculation has been demonstrated, then it stands on its own merits, and its usefulness becomes another question altogether. In fact, De Broglie (1924, p. 33), speaking in a slightly different context, gives us a criterion for this usefulness:

"This hypothesis ... is worth what every hypothesis is, that is, as much as the consequences which can be deduced from it."

However, the calculational avenue of approach, rather surprisingly, is not as straightforward as it might at first appear. It depends to some extent on the assumptions we are prepared to allow. For example, most physicists (including the present author) are not willing to tolerate a model in which a material mass is required to move faster than the velocity of light, and many electron models have been rejected on this basis. In particular, if we adopt the classical electron radius, $R_0 = e^2/mc^2 = 2.8 \times 10^{-13}$ cm, which is the electron radius that was most frequently invoked during the first quarter of this century (Kramers,

1958, p. 233), both the spin angular momentum and the magnetic moment of the electron classically mandate circumferential velocities in excess of c. Hence the assertion was frequently made that classical representations of the electron are impossible.

A crucial question that arises in connection with electron models has to do with the nature of the electron mass. As we have discussed, a purely electrostatic mass is intrinsically unstable. This is a direct consequence of Earnshaw's theorem, and the addition of magnetic fields does not solve the problem (Oppenheimer, 1970, p. 86). Furthermore, Casimir forces, which arise from the reaction of the vacuum state on the particle, do not provide the necessary confining field; in fact, they act as a de-confining mechanism (Boyer, 1968, 1969). Thus we are forced, from stability considerations alone, to introduce a non-electromagnetic force that holds the electron together. If we were to consider an extremely small size for the electron, $R \sim 10^{-33}$ cm, then gravitational forces could be invoked to solve the stability problem. But if we confine ourselves to distance scales of 10^{-17} cm or greater, as we do here, then the gravitational forces are negligible, and we are left with "mechanical" forces, and hence a "mechanical" mass, as our only remaining solution to the confinement problem.

This conclusion is buttressed by energy considerations. We pointed out in Chapter 7 that the intrinsic self-energy W_E of the electric charge e, from a semiclassical viewpoint, must necessarily be equal to zero. And we demonstrated in Chapter 8 by direct calculation that the magnetic self-energy W_H of a Compton-sized electron, while non-zero, amounts to only about 0.1% of the total electron mass. Thus the bulk of the mass must come from some other source than electromagnetism. Having ruled out gravity, we again arrive at a *mechanical* mass as our remaining alternative, where the label *mechanical* is almost a generic term that applies to what is left of the electron after its electromagnetic and gravitational components have been removed. This mechanical mass is our pragmatic answer to the problem posed by Pais (1982, p. 159) in Chapter 1:

"...we still do not know what causes the electron to weigh."

The question as to the nature of the electron mass may actually be one of the strongest arguments in favor of spectroscopic particle models. If we attempt to construct such a model for the electron, we are forced to assign it a mechanical mass. And we are then obliged to determine the properties that must be assigned to this mass in order to account for the experimental data on electrons. In fact, this may be the best avenue of approach we have in deducing these properties, since this mechanical mass does not occur in macroscopic quantities. The assumption of a point particle, which invalidates all of our macroscopic knowledge about mechanics and about electromagnetism as applied to the electron, has discouraged exploration along these lines. But it is of interest to see how far we can proceed in constructing a spectroscopic model, even if just for the sake of pedagogical completeness. This electron model would seem to be a necessary component of the mosaic that we refer to as "modern day physics".

What are the spectroscopic properties of the electron? Its *static* properties are its mass m, electric charge e, and electric dipole moment d. Its *dynamical* properties are its spin angular momentum J, magnetic moment μ, and lifetime τ. The experimental values

TABLE 9.1. SPECTROSCOPIC PROPERTIES OF THE ELECTRON

	static properties
mass m	$0.51099906 \pm 0.00000015$ MeV/c^2
charge e	$(4.8032068 \pm 0.0000015) \times 10^{-10}$ esu
electric dipole moment d	$(-0.3 \pm 0.8) \times 10^{-26}$ e-cm
	dynamical properties
spin angular momentum J	$\hbar/2$
magnetic moment μ	$\hbar/2 \cdot e/mc \cdot (1.001159652193 \pm 0.000000000010)$
lifetime τ	$> 1.9 \times 10^{23}$ years

for these quantities (Particle Data Group, 1992) are summarized in Table 9.1. As can be seen in Table 9.1, the mass, charge, and magnetic moment of the electron are known very precisely. The spin angular momentum is not susceptible to direct quantitative measurement. What is measurable is the gyromagnetic ratio g, which is the ratio of the magnetic moment to the spin (Eq. 8.13). Since the expression for the magnetic moment contains the spin as a factor, the incredibly accurate value quoted for the magnetic moment in Table 9.1 is in actuality the measured value of the gyromagnetic ratio. This is one of the most precise measurements in all of physics.[2] Quantum-mechanically, the spin and magnetic moment values given in Table 9.1 represent the projections along the quantization axes (which are the observed quantities), and the total length of each of these vectors is $\sqrt{3}$ larger (as specified for a spin 1/2 particle).

It is of interest to compare the spectroscopic properties of the electron to those of the muon (Particle Data Group, 1992), which are given in Table 9.2.

TABLE 9.2. SPECTROSCOPIC PROPERTIES OF THE MUON

	static properties
mass m	105.658389 ± 0.000034 MeV/c^2
charge e	$(4.8032068 \pm 0.0000015) \times 10^{-10}$ esu
electric dipole moment d	$(3.7 \pm 3.4) \times 10^{-19}$ e-cm
	dynamical properties
spin angular momentum J	$\hbar/2$
magnetic moment μ	$\hbar/2 \cdot e/mc \cdot (1.001165923 \pm 0.000000008)$
lifetime τ	$(2.19703 \pm 0.00004) \times 10^{-6}$ sec

A comparison of Tables 9.1 and 9.2 demonstrates in a very compelling manner the close relationship between the electron and the muon. The muon appears spectroscopically in all respects as nothing more than a "heavy electron" – an electron whose mass has been increased by a factor of 207, and whose magnetic moment has been reduced by this same factor.[3] The most striking feature of Tables 9.1 and 9.2 is the extreme accuracy of the *measured* magnetic moments (or, more precisely, the *measured* gyromagnetic ratios).

However, even more striking is the fact that the *calculated* values for these quantities are just as accurate as the measured values.[4] This represents a stunning success for QED, and it demonstrates beyond doubt that QED is *the* theory for these particles – or at least for their electromagnetic properties. However, this also presents a challenge for QED. If QED is a complete theory for electrons and muons, we might logically expect the muon-to-electron mass ratio to emerge from a QED-type calculation, which is a result that has not yet been achieved. If such a calculation cannot be carried out, it may be an indication that QED is strictly an electromagnetic formalism, and that the mass values follow from a quite different type of systematics (Mac Gregor, 1990). A more modest challenge for QED might be to simply account for the fact that there is a muon: why does it exist?

The existence of a finite electric dipole moment for either electrons or muons is forbidden by both time reversal invariance and parity invariance (Particle Data Group, 1992), and the experimental limits for this quantity are very small, as shown in Tables 9.1 and 9.2. The electron, the lowest-mass charged particle, is believed to be absolutely stable, since it has no lower state into which it can decay. Its lifetime has been determined experimentally to be more than 10^{23} years, far longer than the age of the universe.

The spectroscopic properties shown in Table 9.1 are the features that are to be reproduced by a representational model of the electron. The early models of Abraham, Lorentz, and Poincaré, which were discussed in Chapter 3, dealt mainly with the static properties of the electron. A large number of papers on electron models have subsequently appeared.[5] Many of these center on the point electron, and are in general of a rather abstract nature. None of these papers feature the relativistically spinning sphere, which is the model that, in the opinion of the present author, most naturally ties together all of the information shown in Table 9.1. We now proceed to the mathematical development of the relativistically spinning sphere model of the electron, and a discussion of some of its consequences.

Notes

1 The quotation by A. Einstein shown on page 69 is from Wheeler (1991).

2 Feynman (1985, p. 7) has pointed out that the accuracy of the magnetic moment value given in Table 9.1 is comparable to a measurement of the distance from Los Angeles to New York to an accuracy of the width of a human hair.

3 It is amusing to note that the muon-to-electron mass ratio of 207 is equal to e^2c^2 in cgs units. However, since the mass ratio is dimensionless, whereas e^2c^2 is in units of gm cm^5/sec^4, we are faced, if we take this result seriously, with the problem of deciding why the Creator would attach such fundamental significance to the cgs system, and to these three independent and arbitrarily selected units – grams, centimeters, and seconds.

4 For the calculation of the electron anomalous magnetic moment and its comparison to experiment, see Kinoshita and Linquist (1981); Rich and Wesley (1972). For similar muon references, see Kinoshita, Nizic and Okamoto (1984); Calmet, Narison, Perottet and de Refael (1977).

5 The following is a representative list of papers that deal with various aspects of electron models: Barut and Bracken (1981), Barut and Zanghi (1984), Barut and Unal (1989a), Barut and Dowling (1989b), Bialynicki-Birula (1982), Blanco, Pesquera and Jiminez (1986), Blanco (1987), Bohm and Weinstein

(1948), Boyer (1982, 1985a), Bunge (1955), Caldirola (1956), Caldirola, Casati and Prosperetti (1978), Carmeli (1984a, 1984b, 1985), Cvijanovich and Vigier (1977), Daboul and Jensen (1973), Daboul (1975), de la Peña, Jimenez and Montemayor (1982), Duval (1975), Erber (1961), Franca, Marques and da Silva (1978), Fryberger (1975), Grandy and Aghazadeh (1982), Head and Moore-head (1981), Kaup (1966), Levine, Moniz and Sharp (1977), Lopez (1984), Mannheim (1978), Moniz and Sharp (1977), Nodvik (1964), Nyborg (1962), Pais (n.d.), Pearle (1977), Rafanelli and Schiller (1964), Ranada and Vasquez (1984), Raskin (1978), Rohrlich (1982), Teitelboim, Villarroel and Van Weert (1980). The following books also have references to electron models: Feynman (1985), Jackson (1962), Jauch and Rohrlich (1976), Kramers (1958), Lorentz (1962), Pais (1982), Pearle (1982), Rohrlich (1965, Ch. 2). For review articles, see Miller (1976), Moyer (1981), Pais (1972, pp. 79-93), Rohrlich (1973, pp. 331-369).

Chapter 10

Spin Quantization and the Relativistically Spinning Sphere

We now arrive at the task of constructing a spectroscopic model of the electron. This will occupy the remaining five chapters of this section. We start in the present chapter with two central problems of electron structure as seen from the point of view of classical (relativistic) mechanics: namely, *(1) the relationship between the mass and the spin angular momentum of the electron* (which has interesting ramifications that bear on the question of spin quantization); and *(2)* an even more basic problem, *the nature of the spin angular momentum itself.* Quite surprisingly, the development we give here, which constitutes the most straightforward investigation we can carry out, and which leads directly to the relativistically spinning sphere model, is one that is not to be found anywhere[1] except in papers by the present author.[2]

The fundamental topic we are concerned with in this chapter is the nature of the electron spin. This topic was introduced in Chapter 3, and has been elaborated upon to some extent in subsequent chapters. It remains to this day one of the most arcane subjects in particle physics. Ohanian (1986) describes the current viewpoint of physicists very succinctly:

> "According to the prevailing belief, the spin of the electron or of some other particle is a mysterious internal angular momentum for which no concrete physical picture is available, and for which there is no classical analog."

It is this "prevailing belief" that we seek to undercut with the calculations of the present chapter. The history of the discovery of electron spin has been well documented by van der Waerden (1960), and it is of interest to briefly recapitulate the earliest ideas. The concept of the spinning electron started with an empirical observation by Pauli that an extra two-valued electron degree of freedom is required in order to explain the splitting of atomic energy levels. Although Pauli's ideas suggested that this degree of freedom can be attributed to an electron spin angular momentum, with a corresponding electron magnetic moment, the actual identification of the electron as having these properties was first made by Kronig, and then slightly later and independently by Uhlenbeck and Goudsmit. These were three young physicists just starting out on their careers. Due to the initial opposition this identification engendered, Kronig did not publish his work, and that of Uhlenbeck and Goudsmit was published only because Ehrenfest, their senior adviser, took it upon himself to send to Naturwissenschaften a brief note they had written (Uhlenbeck and Goudsmit, 1925), but which they felt was too speculative to submit. The success of these ideas in explaining atomic spectra was soon apparent, especially after the work of Thomas (1926), so that the spinning electron quickly became an accepted member of the physics particle world. But the nature of the spin itself became a problem that has remained unresolved. The first assumption, as related by Uhlenbeck many years later (van der Waerden, 1960, p. 213), was that the electron must be a rotating sphere:

> "Goudsmit and myself hit upon this idea by studying a paper of Pauli, in which the famous exclusion principle was formulated and in which, for the first time, *four* quantum numbers were ascribed to the electron. This was done rather formally; no concrete picture was connected with it. To us, this was a mystery. We were so conversant with the proposition that every quantum number corresponds to a degree of freedom, and on the other hand with the idea of a point electron, which obviously had three degrees of freedom only, that we could not place the fourth quantum number. We could understand it only if the electron was assumed to be a small sphere that could rotate ... ."

However, the attempts to deduce the properties of this spinning sphere led to difficulties that we have documented in the preceding chapters. The one rotational aspect of the electron that seems firmly established is that it corresponds to a two-valued representation of the three-dimensional rotation group (see Chapter 13). As Pauli himself described the matter (van der Waerden, 1960, p. 216), this is the aspect of the electron spin that has conceptually survived:

> "After a brief period of spiritual and human confusion, caused by a provisional restriction to 'Anschaulichkeit', a general agreement was reached following the substitution of abstract mathematical symbols, as for instance psi, for concrete pictures. Especially the concrete picture of rotation has been replaced by mathematical characteristics of the representations of the group of rotations in three-dimensional space."

This renunciation of *anschaulichkeit*, or visualizability, is still the customary point of view, as was recapitulated many years later by Margenau (1961, p. 6) (see Chapter 3). In discussing these results, van der Waerden (1960, pp. 215-216) commented that

> "spin cannot be described by a classical kinematic model, for such a model can never lead to a two-valued representation of the rotation group".

This statement characterizes the type of reasoning that has generally precluded further investigation along these lines. However, as we describe in Chapter 13, it is possible, by adding in electromagnetic effects, to obtain a "classical kinematic model" that *does* lead to two-component spin wave functions, and hence stands as a counterexample to van der Waerden's comment. In the present chapter we restrict ourselves to the strictly classical aspects of electron spin. These aspects do not pertain to the quantum questions, but they do pertain to both special and general relativity, and in a very interesting manner.

How shall we treat the spin of the electron classically? Guided by the discussions of the previous chapters, we make the following *ansatz*:

The electron mass is a continuous distribution of ***mechanical*** *matter.*

If we start with this *ansatz*, and then apply the standard rules of present-day physics, what can we deduce about this representation for the electron? A considerable amount, as it turns out. Let us treat this problem as it might be handled in a physics lecture on special relativity.

The *first fact* we know about the electron is that it is spinning. This suggests that its mass has rotational symmetry. Quantum-mechanically, the electron spin consists of a rotation around the spin axis (the vector ω in Figure 7.1) combined with a precessional motion (specified by the angle ϕ) around the z-axis of quantization. Hence the rotational symmetry of the mass may be three-dimensional, or spherical. We will ascribe to the electron the simplest spherical shape we can devise: a uniform sphere of matter (in the rest frame of the mass).

The *second fact* we know about the electron is that its spin is quantized: all electrons have the same spin angular momentum. This spin quantization logically occurs for one of two reasons:

(a) the electron is spinning as fast as it is allowed to – that is, with its equator moving at, or infinitesimally below, the velocity of light, c;

(b) the spin energy of the electron is dictated by the energetics of the electron production mechanism, which is isoergic.

As we will see by the end of the chapter, *both* of these factors may be operating here. The assertion is sometimes made that there is no energy associated with the spin of the electron (Schiff, 1955, p. 331). However, this assertion is not only contrary to our intuitive notions, but it also seems to be in disagreement with the fact that electron spin angular momenta couple to photon spin angular momenta, which in turn are capable of producing macroscopic rotations in a half-wave crystalline plate (Cagnac and Pebay-Peruoula, 1971, p. 271).

The *third fact* we know about the electron is that its spin, $J = 1/2\ \hbar$, is very large, especially for an object as light as the electron. (The proton, for example, which is 1800 times as massive as the electron, has the same spin angular momentum as the electron.) This suggests that the electron must be spinning very rapidly, so that its motion is relativistic. Hence the spherical mechanical mass that we have attributed to the electron must exhibit a relativistic increase in mass as compared to its non-spinning value. Since the various mass elements in the sphere move at different instantaneous velocities, depending on their distances from the rotational axis of the sphere, the uniform density that prevailed in the sphere in its rest frame is modified by the relativistic motion: the density of the spinning sphere increases as its equator is approached. This relativistic mass increase changes not only the observed mass of the spinning sphere, but also the ratio between the mass and the moment of inertia of the sphere.

The *fourth fact* we know about the spinning electron is that its spherical *shape* is not changed relativistically, since the rotational motion is parallel to the surface of the sphere. The spherical outline of the electron is relativistically unaffected by the spin motion. This same conclusion holds for axially centered ring elements within the sphere. Einstein described this situation as follows (Einstein, 1923, p. 116), using K and K′ as fixed and rotating coordinate systems, respectively:

"For reasons of symmetry it is clear that a circle around the origin in the X,Y plane of K may at the same time be regarded as a circle in the X′,Y′ plane of K′. ... if we envisage the whole process of measuring from the 'stationary' system K, ... the measuring-rod (in K′) applied to the periphery undergoes a Lorentzian contraction, while the one applied along the radius does not."

More recently, Penrose (1959) has shown that a linearly moving sphere, even though it is flattened relativistically along the line of motion, nevertheless appears to an observer to be spherical when the finite propagation time of the light from the sphere to the observer is taken into account.

Guided by these facts, let us now calculate the relativistic mass increase of a rapidly spinning sphere. To accomplish this, we divide it up mathematically into mass elements which are equidistant from the axis of rotation. The simplest mass element is a ring centered on the rotational axis. If we denote the rest mass of the ring as $m_0(r)$, where r is the distance from the axis, then the relativistic mass increase is

$$m(r) \quad = \quad \frac{m_0(r)}{\sqrt{1-\omega^2 r^2/c^2}} \quad , \qquad (10.1)$$

where ω is the angular velocity of the spinning ring. Interestingly, this mass increase can be pictured in two different ways: *(1)* it arises from *special-relativistic* effects, where the instantaneous linear velocity of a mass element is $v = \omega r$; or *(2)* it arises from *general-relativistic* effects, where the mass increase is a consequence of the gravitational potential associated with the rotational motion (Møller, 1952, p. 318, Eq. 42). Both viewpoints lead to Eq. (10.1).[3] As noted above, the overall shape of the ring is unaffected relativistically. Thus spinning rings can be superimposed to form a spinning disk or cylinder or sphere.

Computationally, the most convenient way to obtain the relativistic mass of a spinning sphere is to use axially centered cylindrical mass elements, as shown in Figure 10.1.

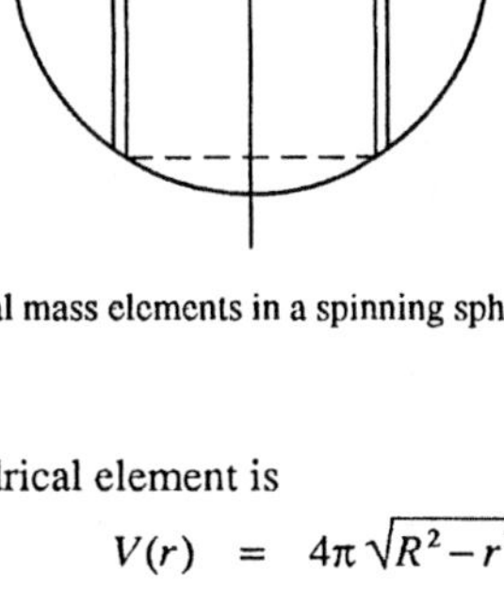

Fig. 10.1. Cylindrical mass elements in a spinning sphere.

The volume of a cylindrical element is

$$V(r) \quad = \quad 4\pi\sqrt{R^2-r^2}\, r\, dr \quad , \qquad (10.2)$$

where R is the radius of the sphere, and r is the distance from the rotational axis. Each cylindrical mass element experiences a relativistic increase that is given by Eq. (10.1). The total mass of the spinning sphere is given by the integral

$$M_S = \frac{3M_0}{R^3} \int_0^R \sqrt{\frac{R^2 - r^2}{1 - \omega^2 r^2/c^2}}\, r dr \quad , \tag{10.3}$$

where M_S and M_0 represent the spinning and non-spinning masses, respectively. For small values of ω, Eq. (10.3) reduces to $M_S \cong M_0$. As ω increases, M_S increases monotonically. The largest value ω can have is

$$\omega = c / R \quad , \tag{10.4}$$

which represents the angular velocity at which the equator of the spinning sphere is moving at the velocity of light, c. When this limiting value for ω is substituted into Eq. (10.3), a remarkable result occurs: the r-dependence in the numerator and denominator of the square root cancels out. As a result, the integral is finite, and it gives

$$M_S = \frac{3}{2} M_0 \quad . \tag{10.5}$$

The relativistic spinning mass M_S is half again as large as the non-spinning rest mass M_0! This is a phenomenologically significant result, as we will see. In the mass integration of Eq. (10.3), the mass density in the cylindrical mass element becomes very large as the rapidly moving equator is approached, but the volume of the cylindrical element decreases by this same factor, so the total mass in the cylindrical element remains finite. This is a non-trivial result, and it follows uniquely from the spherical geometry that we have selected for the electron. Empirically, the velocity at the equator of the sphere is expected to remain infinitesimally below the limiting velocity c, but the corresponding change in the relativistic mass M_S is also infinitesimal. Thus this calculation is mathematically well-behaved.

The above calculation can also be viewed from a somewhat different relativistic perspective. The cylindrical volume elements used in Eqs. (10.2) and (10.3) are based on the implicit assumption that they apply to *Euclidean* space. When we study a cylindrical mass element of the spinning sphere as observed in the non-spinning rest frame (that is, in Einstein's K frame), we find that its total mass is increased relativistically, in accordance with Eq. (10.1). Since the perceived *Euclidean envelope* of the cylinder (Figure 10.1) is unchanged, the density of the mass has apparently increased. However, the *circumference* of the cylindrical element, if it is actually measured by an observer in K, is also increased by this same factor: *the geometry of the cylinder is no longer Euclidean.* Thus the measured volume of the cylindrical element is increased in the same ratio as the mass, so that the *measured density* of the mass remains constant. From this viewpoint, the volume of the spinning sphere is relativistically increased and is filled by the matching relativistic increase in the mass.

Our next task is to calculate the relativistic moment of inertia I of the spinning sphere. This is given by the equation

$$I = \frac{3M_0}{R^3} \int_0^R \sqrt{\frac{R^2 - r^2}{1 - \omega^2 r^2/c^2}}\, r^3 dr \quad . \tag{10.6}$$

In the rotational limit of Eq. (10.4), this becomes

$$I \;=\; \frac{3}{4} M_0 R^2 \;=\; \frac{1}{2} M_S R^2 \quad . \tag{10.7}$$

This relativistic moment of inertia is larger than the corresponding non-relativistic moment of inertia, $I = 2/5\ M_0 R^2$, due to the increase in mass at large distances from the axis of rotation. The relativistic mass increase $M_S = 3/2\ M_0$ that we obtained from Eqs. (10.1) - (10.5) is a very general result which does not depend on the radius R of the sphere. The moment of inertia I, however, does depend on R, which means that we have to assign a value to R in order to obtain a value for I. This can be accomplished by equating the calculated and measured values of the electron spin J. The spin angular momentum of the relativistically spinning sphere is

$$\vec{J} \;=\; I\,\vec{\omega} \quad , \tag{10.8}$$

so that setting $\omega = c/R$ (Eq. 10.4), and using Eq. (10.7), we have

$$J \;=\; \frac{1}{2} M_S R c \quad . \tag{10.9}$$

If we now set R equal to the "spinning-mass Compton radius",

$$R \;=\; \hbar / M_S c \quad , \tag{10.10}$$

we obtain the spin angular momentum

$$J \;=\; \frac{1}{2}\hbar \tag{10.11}$$

as a *directly calculated quantity*. The relativistically spinning sphere model thus ties together the mass, spin angular momentum, and Compton radius of the electron.

This model provides a reason for the observed spin quantization of electrons: they are all spinning as fast as they are allowed to, with their equators moving at, or infinitesimally below, the velocity of light. It may also provide a second reason. As we saw from Eq. (10.5), a relativistically spinning mass is half again as massive as it was at rest. Thus the transition

$$M_0 M_0 M_0 \;\Leftrightarrow\; M_S M_S \tag{10.12}$$

can occur as an isoergic process. And if the two spinning masses are in a $J = 0$ total spin state, then angular momentum is also conserved. We know that electrons are produced in particle-antiparticle pairs. If Eq. (10.12) represents an appropriate pair-wise production mechanism for electrons,[4] then they are *required* to spin at the full relativistic limit in order to conserve energy. From the standpoint of just the electron, the plausibility of this type of mass conversion process is difficult to assess. However, if we move to the hadronic realm, parallels to this situation can be found. In particular, mass triplings are a prominent feature of the hadron mass spectrum (Mac Gregor, 1990). The best example of this is provided by the *thresholds* at which the various types of hadronic states first appear. This fact is illustrated in Figure 10.2, which shows the dominant $\phi(1020) = s\bar{s}, J/\psi(3097) = c\bar{c}$, and $\Upsilon(9460) = b\bar{b}$ meson resonances (Particle Data Group, 1992), with their masses plotted on a linear mass-tripling trajectory that is normalized to the mass of the Υ. When this linear trajectory is extended upwards in mass, it passes, after two more mass triplings, through the mass centroid of the ultraheavy $W(80600)$ and $Z(91161)$ gauge bosons, which are supposedly unrelated. Furthermore, the hadronic decay cascades of these particles exhibit frequent transitions between spinless and spinning states, so that a process akin to Eq. (10.12) may well be occurring. Figure 10.2 also includes the thresholds for muon pairs and for proton pairs, shown plotted on a second linear mass-tripling trajectory. Although these two particles supposedly belong to different families – leptons and hadrons – their masses are accurately separated by a factor of nine.

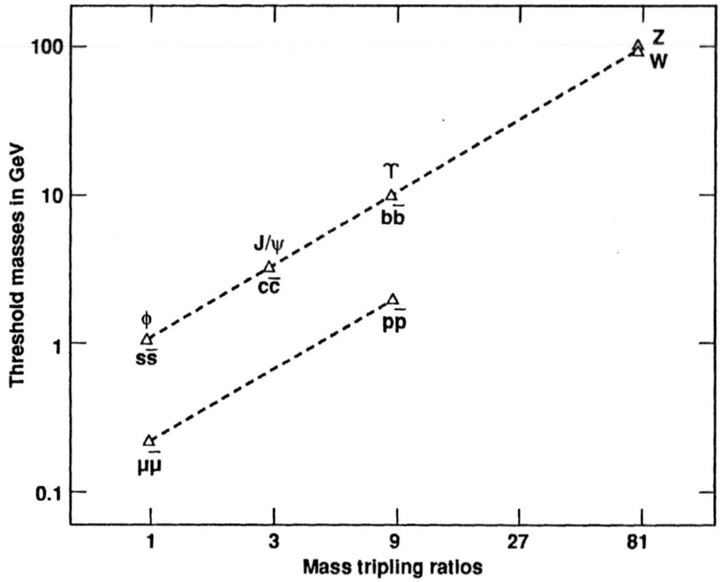

Fig. 10.2. Thresholds for the production of various types of elementary-particle states, showing the very accurate mass-tripling ratios that occur.

What have we accomplished in this chapter? We started with the *ansatz* that the electron is formed as a uniform sphere of *mechanical* matter. Then, by applying the basic equations of mechanics as modified by special relativity, we calculated the spin energy and the moment of inertia of the relativistically spinning sphere. Finally, assigning the Compton radius to the sphere, and assuming that the sphere is spinning at the (quantized) relativistic limit, we calculated the correct spin angular momentum for the electron. The equations that we obtained are very simple ones, and would seemingly be found in some of the papers on electron models.[1] But this is not the path that has been followed. The preconception of the electron as a *point-like* object has so thoroughly dominated present-day particle physics that alternative viewpoints have scarcely been explored. In carrying out the present extension to a *finite-sized* electron, we have obtained two crucial results:

(a) The spin motion of the electron must be calculated relativistically;

(b) When this is done, the moment of inertia of the electron turns out to be

$$I = \frac{1}{2} m r^2 \quad , \tag{10.13}$$

where m is the *observed mass* of the electron, and r is the radius. Eq. (10.13) is the key equation that ties together all of the main spectroscopic properties of the electron, and it was in fact a search for this equation that led the present author to the spinning sphere model.[2] In particular, as we will see in the next chapter, this equation enables us to classically reproduce the gyromagnetic ratio of the electron, which is an achievement that is often regarded as being impossible.[5]

As a final result in this chapter, we note that the correct relationship between the mass, spin angular momentum, and Compton radius of the electron was obtained from a relativistic calculation in which no distortional effects were included. A factor which is possibly operating here is that the mechanical mass of the electron functions effectively as a "rigid body", in which all internal stresses are "instantaneously" distributed throughout the sphere. This may seem at first glance to be a rather implausible assumption. However, it must be kept in mind that the macroscopic "mechanical" masses with which we are familiar are in reality electromagnetically bound aggregates of atoms that are composed mostly of empty space (see Chapter 15). We have no macroscopic representation of a true "rigid body", and the properties that an electron "mechanical" mass might possess may well include that of "rigidity". As additional support for this conclusion, we note that the mass integrations of Eqs. (10.3) and (10.6) are based on the implicit assumption that the mass distribution inside the spinning sphere is continuous (or at least granular on a scale that is much smaller than the Compton radius of the electron). From this viewpoint, "electron matter" is not mostly "empty space", but instead *represents a continuous distribution function which has no counterpart in the macroscopic world.*

As we have just seen, the relativistically spinning sphere model properly ties together the main *mechanical* features of the electron. In order to see how it handles the *electromagnetic* features, we must add in the electric charge. This is the subject of the next chapter.

Notes

1 The author has reviewed the following textbooks and treatises on special relativity, and none of these make any mention of the relativistically spinning sphere model: Einstein *et al.* (1923), Dingle (1940), Bergmann (1942), Møller (1952), Cullwick (1958), Pauli (1958), Stephenson and Kilminster (1958), Coleman (1958), Born (1962), Bridgman (1962), Fock (1964), Smith (1965), Bohm (1965), Aharoni (1965), Synge (1965), Ray (1965), Kilmister (1965), Rindler (1966), Tonnelat (1966), Arzelies (1966), Costa de Beauregard (1966), Kacser (1967), Guggenheim (1967), Prokhovnik (1967), Rosser (1967), French (1968), Mermin (1968), Rosser (1968), Terletskii (1968), Shadowitz (1968), Reichenbach (1969), Janossy (1971), Essen (1971), Taylor (1975), Rindler (1977), Miller (1981), Hoffmann (1983), Goldberg (1984). The following books contains discussions of aspects of special relativity: Mercier and Kervaire

(1956), Oppenheimer (1964), Russell (1969), Lucas (1973). There have been a large number of papers published on relativistically spinning disks and spheres, of which we cite a few examples: Hill (1946, 1947), Rosen (1946, 1947), Kursunoglu (1951), Takeno (1952), Gardner (1952), Synge (1952), Hogarth and McCrea (1952), Salzman and Taub (1954), Dewan and Beran (1959), Phipps (1962a, 1962b), Boyer (1962), Lichtenberg (1962), Fletcher (1963), Dewan (1963), Grøn (1979), McFarlane *et al.* (1980).

2 Discussions of the relativistically spinning sphere are contained in the following publications by the present author: Mac Gregor (1970), (1974, App. B), (1978, Ch. 6), (1981), (1985a), (1989), (1990, App. A and B).

3 See Møller (1952, pp. 220-221). This is an example of Einstein's Principle of Equivalence. The special-relativistic mass increase represents an inertial mass, and the general-relativistic mass increase represents a gravitational mass. As shown in experiments of the Eötvös type, these masses are equal to one another.

4 Eq. (10.12) obviously does not apply in a direct manner to the production of single electrons via nuclear beta decay.

5 See Note 4 in Chapter 2.

Chapter 11

A Classical Spectroscopic Model of the Electron

In the preceding chapter we developed the equations for the relativistically spinning sphere. This is a sphere of uniform *mechanical* matter which is set into motion and spun at the relativistic limit, with its equator moving at (or just below) the velocity of light c. The relativistic sphere equations led to the following relationships:

$$m_S = \frac{3}{2} m_O , \qquad (s, \mathrm{o} = \text{spinning, non-spinning}) ; \tag{11.1}$$

$$I = \frac{1}{2} m_S R^2 , \qquad (\mathrm{R} = \text{radius of the sphere}) ; \tag{11.2}$$

$$J = I \omega = \frac{1}{2} m_S R c , \qquad (\omega \cong c / R) ; \tag{11.3a}$$

$$J = \frac{1}{2} \hbar , \qquad (\omega \cong c / R ; \quad R = \hbar / m_S c) . \tag{11.3b}$$

These relationships give the correct classical value for the spin angular momentum of the electron. In order to make this sphere into a *model* for the electron, we have to add in electromagnetic effects. That is, we must place a charge e on the relativistically spinning sphere. Our goal here is to reproduce the gyromagnetic ratio g of the electron,

$$g = \frac{\mu}{J} \cdot \frac{2 m_S c}{e} , \tag{11.4}$$

which is determined by the charge and mass distributions in the electron. The most convenient way to characterize these distributions is in terms of root mean square (rms) charge and mass radii (Singh and Raghuvanshi, 1984). The magnetic moment vector that corresponds to a *rotating charge distribution* in the electron is

$$\vec{\mu} = \frac{1}{2c} \int \vec{r} \times (\vec{\omega} \times \vec{r}) \, \rho_c(\vec{r}) \, dV , \tag{11.5a}$$

where $\rho_c(\vec{r})$ is the local charge density in esu. The angular momentum vector that corresponds to a *rotating mass distribution* in the electron is

$$\vec{J} \;=\; \int \vec{r}\times(\vec{\omega}\times\vec{r})\,\rho_m(\vec{r})\,dV\,, \qquad (11.5b)$$

where $\rho_m(\vec{r})$ is the local mass density. The total charge and mass values are

$$e \;=\; \int \rho_c(\vec{r})\,dV \qquad (11.6a)$$

and

$$m_S \;=\; \int \rho_m(\vec{r})\,dV\,, \qquad (11.6b)$$

respectively. In general, $\vec{J}$ and $\vec{\mu}$ are not parallel to $\vec{\omega}$, so that the gyromagnetic ratio g is a tensor quantity. However, if the charge and mass distributions are either spherically symmetric or axially symmetric with respect to the spin axis of the sphere, then all of these vectors are parallel, and g reduces to a scalar. For *spherically symmetric* charge and mass distributions, we have (Singh and Raghuvanshi, 1984)

$$\mu \;=\; \frac{1}{3c}\,\omega\, e < r^2 >_c \qquad (11.7a)$$

and

$$J \;=\; \frac{2}{3}\,\omega\, m_S < r^2 >_m \quad, \qquad (11.7b)$$

where $< r^2 >_c$ and $< r^2 >_m$ are the rms charge and mass radii, respectively, and (r,θ,ϕ) are spherical coordinates. For *axially symmetric* charge and mass distributions, which represent the case of interest here, we have

$$\mu \;=\; \frac{1}{2c}\,\omega\, e < r^2 >_c \qquad (11.8a)$$

and

$$J \;=\; \omega\, m_S < r^2 >_m \quad, \qquad (11.8b)$$

where (r,z,ϕ) are axial coordinates. The equation pairs (11.7a,b) and (11.8a,b) each give

$$\frac{\mu}{J} \;=\; \frac{e}{2m_S c}\cdot\frac{< r^2 >_c}{< r^2 >_m}\;, \qquad (11.9)$$

so that

$$g \;=\; \frac{< r^2 >_c}{< r^2 >_m} \qquad (11.10)$$

for either spherically symmetric or axially symmetric charge and mass distributions. If we now take this equation and apply it to the orbital motion of an electron around an atomic nucleus, we have essentially identical (point-like) charge and mass distributions, so that $< r^2 >_c = < r^2 >_m$, which gives $g = 1$, and which leads to the *normal* Zeeman effect. If we want $g \neq 1$ for the electron itself, so as to obtain the observed *anomalous* Zeeman effect, then we must have $< r^2 >_c \neq < r^2 >_m$. In particular, the electron value $g = 2$ that was postulated by Uhlenbeck and Goudsmit (1926) requires

$$<r^2>_c \;=\; 2<r^2>_m \quad . \tag{11.11}$$

For the case of the relativistically spinning sphere, which represents an axially symmetric mass distribution, Eqs. (11.2), (11.3a), and (11.8b) give

$$I \;=\; \frac{1}{2} m_S R^2 \;=\; m_S <r^2>_m \quad , \tag{11.12}$$

so that

$$<r^2>_m \;=\; \frac{1}{2} R^2 \quad . \tag{11.13}$$

Inserting (11.13) into (11.11) gives

$$<r^2>_c \;=\; R^2 \quad . \tag{11.14}$$

Hence, in order to obtain the gyromagnetic ratio $g = 2$ for the relativistically spinning sphere, the charge e must be placed on the *equator* of the sphere! As far as the calculation of the magnetic moment is concerned, this charge could be either a single point or a distributed ring. Electron scattering experiments, however, mandate a point charge configuration (see Chapter 7). Also, the fact that we require the self-energy W_E of the electric charge to be zero (Chapter 7) indicates that the charge is point-like, since a distributed charge would classically interact with itself to produce a non-zero value for W_E. A point charge, of course, does not represent an axially symmetric charge distribution, but it is axially symmetric when averaged over a cycle of rotation of the electron.

As we have just seen, we can use the gyromagnetic ratio $g = 2$ of the electron to deduce that its charge e must be located on the equator of the spinning sphere. (This of course is where the charge will go if it is free to move about inside the confining mechanical sphere, since the field lines of the induced magnetic field repel one another.) We can also use a direct calculation of the magnetic moment to arrive at this same conclusion, as we now demonstrate. A rotating charge, when averaged over a cycle of rotation, represents a current loop. The (average) magnetic dipole moment that corresponds to this current loop is

$$\mu \;=\; \pi R^2 \cdot i \quad , \tag{11.15}$$

where R is the radius of the loop and i is the current. In cgs units, we have

$$i \;=\; \frac{e}{c} \cdot \frac{\omega}{2\pi} \quad , \tag{11.16}$$

so that if we go to the relativistic limit $\omega = c/R$, we obtain

$$\mu \;=\; \frac{eR}{2} \quad . \tag{11.17}$$

But we know that

$$\mu \;=\; \frac{e\hbar}{2m_S c} \tag{11.18}$$

for the electron. Equating (11.17) and (11.18), we arrive at the charge radius

$$R \;=\; \frac{\hbar}{m_S c} \quad . \tag{11.19}$$

This value for R is equal to the sphere radius (Eq. 10.10) that we were led to in the last chapter . Thus the charge distribution on the electron is equatorial. This result, when combined with the relativistic moment of inertia (Eq. 11.2), enables us to obtain the gyromagnetic ratio $g = 2$ of the electron directly from classical considerations.[1]

The "classical" electron model that we have just obtained consists of a mechanical sphere of radius $R = \hbar/mc$, spinning at the relativistic limit $\omega = c/R$, and containing an equatorial charge e. This is a first-order model, which, while heuristically quite useful, is deficient in several respects. We can illustrate these deficiencies by comparing the calculated spectroscopic properties of this model with the electron values listed in Table 9.1. The mass m of the model is an input parameter, and is set equal to the experimental value shown in Table 9.1. However, this measured mass actually consists of the mechanical mass plus the magnetic self-energy of the spinning sphere. As we demonstrated in Chapter 8, the magnetic self-energy amounts to a factor of $\alpha/2\pi$, or about 1/860, of the total electron mass. Since the magnetic field has zero angular momentum, it does not contribute to the spin of the electron. Thus the calculated electron spin J, if we take this correction into account, is too small by a factor of approximately $(1 - \alpha/2\pi)$. Also, the calculated magnetic moment, $\mu = e\hbar/2mc$, is too small by this same factor. In order to properly reproduce J and μ, we should have taken the sphere radius R as

$$R_m = \frac{\hbar}{m_m c} , \qquad (11.20)$$

where m_m is just the mechanical component of the mass. This phenomenological refinement is included in the "quantum mechanical" electron model described in Chapter 14. The electric charge e is another input parameter, and its location on the equator of the sphere insures that the electric dipole moment d is equal to zero. The final entry in Table 9.1, the stability of the electron, is a property that is beyond the scope of the present monograph.

This first-order electron model thus starts with the observed mass m and charge e as input parameters, uses the sphere radius $R = \hbar/mc$, employs relativistic corrections to the spinning mass, and approximately reproduces the observed spin J and magnetic moment μ. Since these calculated quantities are both low by the same factor of $(1 - \alpha/2\pi)$, the model reproduces the gyromagnetic ratio g correctly to first order in α. It also reproduces the vanishing electric dipole moment. But it has two significant shortcomings: ***(a)*** the spin axis of the electron, which is here assumed to coincide with the z-axis of quantization, should actually be inclined at an angle

$$\theta_{QM} = \pm \arccos 1/\sqrt{3} , \qquad (11.21)$$

and the model should produce calculated spin and magnetic moment values which are each $\sqrt{3}$ larger, so that the *projections* of these quantities along the z-axis correspond to the values shown in Table 9.1; ***(b)*** the equatorial charge distribution on the spinning sphere gives rise to an electric quadrupole moment, which increases the electric "size" of the electron over that of a point particle, and which would be observed (and is not) in bound-state electron orbitals. As we have already demonstrated in Chapter 7 (see the discussion of *quantum current loops*), when we remedy the first of these shortcomings, we simultaneously remedy the second. These deficiencies are dealt with in the "quantum mechanical" electron model of Chapter 14.

The rapidly spinning electron that we have described here has several interesting relativistic features. The most important of these from the present viewpoint is the relativistic mass increase and the concomitant increase in the moment of inertia. But there are other effects, some of which were mentioned in Chapter 10. The relativistic characteristics of rotating coordinate systems were described in an early paper by Einstein. We have already quoted briefly from this paper in Chapter 10. It is of interest to repeat this quotation here in more detail (Einstein, 1923, pp. 115-116):

> "In classical mechanics, as well as in the special theory of relativity, the co-ordinates of space and time have a direct physical meaning. To say that a point-event has the X_1 co-ordinate x_1 means that the projection of the point-event on the axis of X_1, determined by rigid rods and in accordance with the rules of Euclidean geometry, is obtained by measuring off a given rod (the unit of length) x_1 times from the origin of co-ordinates along the axis of X_1. To say that a point-event has the X_4 co-ordinate $x_4 = t$, means that a standard clock, made to measure time in a definite unit period, and which is stationary relatively to the system of co-ordinates and practically coincident in space with the point-event, will have measured off $x_4 = t$ periods at the occurrence of the event.
>
> "This view of space and time has always been in the minds of physicists .... This is clear from the part which these concepts play in physical measurements .... But we shall now show that we must put it aside and replace it by a more general view ....
>
> "In a space which is free of gravitational fields we introduce a Galilean system of reference $K(x,y,z,t)$, and also a system of co-ordinates $K'(x',y',z',t')$ in uniform rotation relatively to K. Let the origins of both systems, as well as their axes of Z, permanently coincide. We shall show that for a space-time measurement in the system K' the above definition of the physical meaning of lengths and times cannot be maintained. For reasons of symmetry it is clear that a circle around the origin in the X,Y plane of K may at the same time be regarded as a circle in the X',Y' plane of K'. We suppose that the circumference and diameter of this circle have been measured with a unit measure infinitely small compared with the radius, and that we have the quotient of the two results. If this experiment were performed with a measuring-rod at rest relatively to the Galilean system K, the quotient would be π. With a measuring-rod at rest relatively to K', the quotient would be greater than π. This is readily understood if we envisage the whole process of measuring from the "stationary" system K, and take into consideration that the measuring-rod applied to the periphery undergoes a Lorentzian contraction, while the one applied along the radius does not. Hence Euclidean geometry does not apply to K'. The notion of co-ordinates defined above, which presupposes the validity of Euclidean geometry, therefore breaks down in relation to the system K'. So, too, we are unable to introduce a time corresponding to physical requirements in K', indicated by clocks at rest relatively to K'. To convince ourselves of this impossibility, let us imagine two clocks of identical constitution placed, one at the origin of co-ordinates, and the other at the circumference of the circle, and both envisaged from the "stationary" system K. By a familiar result of the special theory of relativity, the clock at the circumference – judged from K – goes more slowly than the other, because the former is in motion and the latter at rest. An observer at the common origin of co-ordinates, capable of observing the clock at the circumference by means of light, would therefore see it lagging behind the clock beside him. ... So he will be obliged to define time in such a way that the rate of a clock depends upon where the clock may be."

This discussion of Einstein's contains a number of relevant points. As we noted in Chapter 10, a circle in K' keeps its same shape in K. However, the ratio of the circumference C to the radius r is not the same. In K we have $C = 2\pi r$. But an observer in K, using measuring rods in K', sees $C' > 2\pi r'$. Thus the moving primed co-ordinates project into the stationary K frame in a non-Euclidean manner. The instantaneous velocity at r', as observed in K, is $v' = \omega r'$, and the observed circumference is $C' = 2\pi r'\gamma'$, where $\gamma' = 1/\sqrt{1 - v'^2/c'^2}$. Also, a clock placed at r' is observed in K to be running too slowly by a factor of γ'.

It is instructive to calculate the motion of a revolving point as measured with both primed and unprimed coordinates. Let us select a reference radial coordinate r, and a moving coordinate r' that coincides with r at time t and at the spatial position $x = x' = 0$. The relative linear velocity between these points is

$$\mathrm{v} = \frac{dx}{dt} = \frac{\gamma' dx'}{\gamma' dt'} = \mathrm{v}' , \qquad (11.22)$$

which means that $\gamma = \gamma'$. The angular velocity is

$$\omega = \frac{\mathrm{v}}{r} = \frac{\mathrm{v}'}{r'} = \omega' . \qquad (11.23)$$

The radial acceleration is

$$a = \mathrm{v}\frac{d\phi}{dt} = \mathrm{v}\frac{dx}{r}\frac{1}{dt} = \mathrm{v}'\frac{\gamma dx'}{r'}\frac{1}{\gamma dt'} = a' . \qquad (11.24)$$

These results hold for all values of $r \leq R$. An oscillator placed at the equator R, where the velocity is $\mathrm{v}' = \mathrm{c}$, seems to an observer in K to have zero frequency, $f' = 0$. Thus if we apply the de Broglie frequency relation for material particles, $f' = \mathrm{E}'/\mathrm{h}$, to this oscillator, it appears in K with zero energy, $\mathrm{E}' = 0$. In this connection, it is perhaps pertinent to note that the electron charge e, which in the present model resides on the equator of the electron, necessarily has self-energy $W_E = 0$, as we discussed in Chapter 7. Several other workers have also come to the conclusion that $W_E = 0$, as we mention at the end of Chapter 15.

In this model of an extended electron with an equatorial point charge, the charge e follows a circular orbit at the velocity of light c, thus generating the magnetic moment of the electron. Why doesn't the charge e radiate, since it is undergoing a radial acceleration? In the unprimed coordinate system, the radius of curvature of the electron trajectory is measured to be vanishingly small, since the circumference of the path is stretched out relativistically whereas the radius is unchanged. However, the time coordinate is also stretched out, so the acceleration of the charge e is invariant (Eq. 11.24). Thus the charge, according to Maxwell's equations, should radiate. But it clearly doesn't. The answer must lie in the nature of the electron wave that corresponds to this motion. An electron that is bound to an atomic nucleus radiates until it attains a stable orbit in which its associated wave becomes a standing wave. The wave in some manner suppresses the radiation. Similarly, the wave motion that we would associate with a stationary but spinning electron must be such as to preclude radiation. This of course is not a theory, but rather an opinion. However, it may be mandated by the realities of the situation.

If we further extend the ramifications of this spinning sphere model, we encounter one of the pitfalls that can occur in phenomenological physics: *the fact that the correct answer is obtained does not necessarily mean that the calculation is meaningful.* This pitfall has to do with the de Broglie wave equation,

$$\lambda = \frac{h}{p} . \qquad (11.25)$$

In deriving this equation, de Broglie assumed that each particle is associated with "a fictitious wave" ("une onde fictive") (Jammer, 1966, p. 243) that has an angular frequency

$$\omega = \frac{E}{\hbar} = \frac{mc^2}{\hbar} \tag{11.26}$$

in the rest frame of the particle. This angular frequency emerges naturally from the present "classical" electron model. If we set $R = \hbar/mc = c/\omega$ (Eqs. 10.10 and 10.4), we obtain $\hbar\omega = mc^2$, which is just Eq. (11.26). Thus this model provides a natural "explanation" for the de Broglie wave frequency: it is equal to the rotational frequency of the spinning sphere. However, this explanation runs into immediate difficulties. The above equations apply to the situation when the center of the spinning sphere is at rest. If the sphere is moving, then the mass m is increased relativistically, so that the wave frequency ω *increases* (Eq. 11.26). But the sphere itself, which functions as a "clock" to an observer, appears to *rotate more slowly*. De Broglie demonstrated that the sphere in fact stays in phase with the wave, but these two frequencies are clearly different when the center-of-mass of the sphere is moving relative to an observer. Moreover, when we make the alterations to the spinning sphere that are required in the "quantum mechanical" model of Chapter 14, wherein the sphere radius is increased by a factor of $\sqrt{3}\,(1 + \alpha/2\pi)$, we destroy any relationship between the rotational frequency of the sphere and the frequency of the de Broglie wave. This conclusion is strengthened if we examine composite particles – the negatively charged antiproton, pion, and kaon. All of these particles can be bound into electron-like atomic orbitals, and they all exhibit wave characteristics that accurately reflect the de Broglie wave length of Eq. (11.25). Yet they have complex and varied quark substructures, which involve not only unequal quark masses, but also large (and unknown) quark binding energies, and which defy attempts to relate quark rotational frequencies to de Broglie wave frequencies.[2] And the neutron, another compound particle, also accurately obeys the de Broglie wave equation, as demonstrated in crystal interference experiments.[3] Thus it is empirically clear that the de Broglie frequency of a particle is not related to the rotational frequencies of its constituent parts. The de Broglie particle wave is a wave that (to the extent it can be assigned physical reality) exists in the vacuum state (Mac Gregor, 1988), and its frequency is dictated by the coupling between the particle and the vacuum state – a coupling which depends solely on the overall mass and velocity of the particle.

The above discussion is beyond the scope of the present monograph, and will be dealt with elsewhere.[4] It is included here merely to illustrate the fact that we should not expect the rotational frequency of the spinning sphere to correspond to the de Broglie wave frequency. It also emphasizes the fact that in studying the phenomenology of the electron, we must also keep in mind the phenomenology of the other elementary particles.

We have discussed the relativistically spinning sphere as a model for the electron, where the electron is viewed at rest in the laboratory frame of reference – Einstein's K frame. How does this model of a finite-sized electron transform from one co-ordinate frame to another? In the early development of quantum electrodynamics, Feynman, in particular, pointed out the advantages of having a covariant formalism – one which has the same form in all frames of reference. Since the electron itself transforms relativistically, any model for the electron must exhibit these same properties. As we will see in the next chapter, the relativistically spinning sphere fulfills this requirement.

Notes

1 For discussions of classical calculations of the gyromagnetic ratio g of the electron, see Kramers (1958, p. 233), Rohrlich (1965, pp. 204-207) and Pais (1988, pp. 274-280). None of these discussions involve the concept of the relativistically spinning sphere that is used in the present work.

2 If we devise a quark model in which the pion and kaon are composed of spinless quarks (Mac Gregor, 1990), then there is no internal particle frequency.

3 See for example Klein (1983), Greenberger (1983) and Atwood (1984).

4 Mac Gregor, M. H., *The Enigma of Particles and Waves* (manuscript in progress).

Chapter 12

The Lorentz Invariance of a Finite-Sized Electron

In the last two chapters we have studied relativistic effects as they appear in a *rotating* coordinate system versus a *stationary* one. Specifically, we have made a comparison between a relativistically spinning mechanical mass and its spinless counterpart. In the present chapter we study relativistic invariance as it applies to *translational* motion. The electron follows the translational rules of special relativity, and a model for the electron must do likewise. If we picture the electron in the *center-of-mass frame* (where its center is stationary) as consisting of a spatially extended spinning sphere whose mass components are relativistically modified, then we must demonstrate that this picture carries over to the *laboratory frame* (where its center is moving) in a covariant manner.

What are the primary electron properties whose transformations are to be studied? There are three: the electron mass, spin, and magnetic moment. For an electron moving with velocity v in the laboratory frame, the characteristic relativistic transformation parameter is

$$\gamma = \frac{1}{\sqrt{1 - v^2/c^2}} \quad . \tag{12.1}$$

The center-of-mass to laboratory electron transformation properties are:

$$m_{lab} = \gamma \, m_{cm} \quad . \tag{12.2}$$

$$J_{lab} = J_{cm} \quad . \tag{12.3}$$

$$\mu_{lab} = \mu_{cm} \, / \, \gamma \quad . \tag{12.4}$$

$$g_{lab} = g_{cm} \, / \, \gamma \quad . \tag{12.4'}$$

In the laboratory frame, the observed mass m is increased by a factor of γ, the observed spin J is unchanged, and the observed magnetic moment μ and gyromagnetic ratio g are each decreased by a factor of γ. The increase in mass with increasing velocity (French, 1968, pp. 20-24) and the invariance of the spin at all velocities are familiar relativistic results. The decrease of the measured magnetic moment (and hence the gyromagnetic ratio) with increasing velocity is perhaps not as familiar. The gyromagnetic ratio of the electron is conventionally determined by measuring its precessional motion in a magnetic field. This type of measurement is discussed for example by Cagnac and Pebay-Peyroula, who point out (1975, p. 295) that

"... (for) high-energy electrons moving at relativistic velocities - the value of the gyromagnetic ratio depends on the velocity v of the electrons".

Then, on the following page (1975, p. 296, Comment I), they note that

"everything takes place as if the spins in the laboratory frame had the gyromagnetic ratio" ... (e/mc, where m is the **relativistic** mass).

(The gyromagnetic ratio is defined by these authors as the direct ratio of the magnetic moment to the spin value; *i.e.*, $g = e/mc$.) Relativistic equations for this precessional motion have been worked out by Bargmann, Michel, and Telegdi (1959), and these equations, in the laboratory frame, feature the relativistic electron mass in the denominator. Thus the measured magnetic moment of the electron, which is proportional to its precessional motion, varies inversely with its relativistic mass, as specified in Eq. (12.4).

The Lorentz transformation equations are formulated in such a way that they implicitly apply to point-like bodies whose internal structures can be ignored. How shall we apply them to the spatially extended relativistically spinning sphere, whose internal structure is one of its most important features? Analytically, this poses a difficult problem. But numerically, it has a ready solution. We simply divide the sphere up into incremental mass units, transform each unit separately, and then recombine them. In order to achieve sufficient accuracy, a large number of mass units must be used. We start with a rectangular coordinate set:

$$x, y, z \quad = \quad \text{non-rotating c.m. coordinates} \quad . \tag{12.5}$$

The x-axis is the rotational axis of the sphere. The sphere is then divided into rings, each of which has a square cross section in the x,y plane. For the calculations described here, 384 rings were used, each with 96 azimuthal segments, giving a total of about 37,000 mass elements. Finer zoning did not appreciably improve the results. We next define a rotated coordinate set,

$$\begin{aligned} x' &= x \cos\theta + y \sin\theta \quad , \\ y' &= y \cos\theta - x \sin\theta \quad , \\ z' &= z \quad . \end{aligned} \tag{12.6}$$

The velocity $\vec{v}$ of the electron in the laboratory frame is along the x' axis, so that the electron spin axis is at an angle θ with respect to the velocity vector. The coordinates of the 37,000 sphere mass elements were first calculated in the x, y, z frame, and then transformed into the x', y', z' frame. In the laboratory frame, these coordinates become

$$\begin{aligned} x_{lab} &= x'/\gamma \quad , \\ y_{lab} &= y' \quad , \\ z_{lab} &= z' \quad , \end{aligned} \tag{12.7}$$

due to the relativistic contraction of length. The velocities $\dot{x}, \dot{y}, \dot{z}$ of the 37,000 mass elements of the spinning sphere were similarly calculated in the x, y, z frame, and then transformed into the rotated x', y', z' frame. Finally, in the laboratory frame these velocities become

$$\begin{aligned} \dot{x}_{lab} &= \frac{\dot{x}' + \mathrm{v}}{1 + \dot{x}'\mathrm{v}/c^2} , \\ \dot{y}_{lab} &= \frac{\dot{y}'/\gamma}{1 + \dot{x}'\mathrm{v}/c^2} , \\ \dot{z}_{lab} &= \frac{\dot{z}'/\gamma}{1 + \dot{x}'\mathrm{v}/c^2} , \end{aligned} \tag{12.8}$$

where the relativistic addition of velocities is used (French, 1968, p. 126). These equations enable us to calculate the relativistic mass and spin angular momentum of each transformed mass element, and hence of the transformed spinning sphere. For the calculation of the magnetic moment, a series of fractional point charges were placed on the equator of the sphere, so as to represent an average over the rotational motion of a single point charge. The relativistic velocity and coordinate transformations then lead to the transformed value of the magnetic moment.[1]

The results of these calculations are summarized in Table 12.1.

TABLE 12.1. CALCULATED SPECTROSCOPIC TRANSFORMATION PROPERTIES OF THE ELECTRON

Lab frame values as percentage changes from c.m. values. The electron velocity v/c = 0.3.

Angle θ	*Mass*	*Spin*	*Mag. mom.*
0°	+4.80%	+0.00%	-4.61%
30°	+4.81%	+0.04%	-4.68%
60°	+4.81%	+0.10%	-4.78%
90°	+4.82%	+0.12%	-4.82%
"Experiment"	+4.83%	+0.00%	-4.61%

The "experimental" values shown in Table 12.1 are obtained from Eqs. (12.2 - 12.4), and are independent of the polarization angle θ. As can be seen in Table 12.1, the *calculated* transformed values for *m*, *J* and μ are in good agreement with the corresponding *"experimental"* values, thus demonstrating that this electron model has the proper Lorentz transformation properties.

We can take advantage of the computational features of these Lorentz transformations to study them in more detail. Specifically, we can separate out the relativistic *coordinate* transformations (Eq. 12.7) from the relativistic *velocity* transformations (Eq. 12.8), and exhibit each component separately. This is displayed in Table 12.2, which is an expanded version of the example shown in Table 12.1.

TABLE 12.2. SEPARATED COORDINATE AND VELOCITY RELATIVISTIC TRANSFORMATIONS ($\beta = 0.3$)

Lab frame values as percentage changes from c.m. values. The electron velocity $v/c = 0.3$.

Angle θ	*Rel. component*	*Mass*	*Spin*	*Mag. mom.*
0°	coordinate	+0.00%	+0.00%	+0.00%
0°	velocity	+4.80%	+0.00%	-4.61%
0°	both	+4.80%	+0.00%	-4.61%
30°	coordinate	+0.00%	-0.58%	-0.58%
30°	velocity	+4.81%	+0.62%	-4.12%
30°	both	+4.81%	+0.04%	-4.68%
60°	coordinate	+0.00%	-1.73%	-1.73%
60°	velocity	+4.81%	+1.83%	-3.11%
60°	both	+4.81%	+0.10%	-4.78%
90°	coordinate	+0.00%	-2.30%	-2.30%
90°	velocity	+4.82%	+2.43%	-2.57%
90°	both	+4.82%	+0.12%	-4.82%

Table 12.2 shows an interesting breakdown of the relativistic corrections. The *coordinate* corrections (contractions of length) do not affect the mass values, and they affect the spin and magnetic moment values in precisely the same manner. The *velocity* corrections, on the other hand, which are totally responsible for the relativistic mass values, exhibit a marked dependence on the angle θ, and they appear in the spin and magnetic moment corrections with opposite signs. In the spin corrections, the coordinate and velocity contributions essentially cancel out, so that the spin value remains unchanged. In the magnetic moment corrections, however, the coordinate and velocity contributions have the same (negative) sign, and they combine together in an angle-dependent manner to produce the proper (angle-independent) relativistic magnetic moment correction.

The manner in which these results extend to other electron velocities is illustrated in Table 12.3. The "experimental" values shown in Table 12.3 are from Eqs. (12.2 - 12.4). The agreement between "experiment" and calculations is not perfect, but it is close enough to show that the relativistic systematics is essentially being reproduced. This detailed examination of the relativistic corrections that apply to a spatially extended spinning sphere appears to be an original result (Mac Gregor, 1985), and it illustrates the usefulness of having large-scale computational facilities available for numerical calculations.[2]

The above results apply to the "classical" electron model of Chapter 11, which has its spin axis directed along the axis of quantization. In the "quantum mechanical" model of Chapter 14, the spin axis is inclined at an angle of 54.7° to the axis of quantization

TABLE 12.3. SEPARATED COORDINATE AND VELOCITY RELATIVISTIC TRANSFORMATIONS ($\beta = 0.1$ and 0.7)

Angle θ	*Rel. component*	*Mass*	*Spin*	*Mag. mom.*
		Relativistic electron velocity $v/c = 0.1$.		
60°	coordinate	+0.00%	-0.19%	-0.19%
60°	velocity	+0.49%	+0.20%	-0.33%
60°	both	+0.49%	+0.02%	-0.52%
"Experiment"		+0.50%	+0.00%	-0.50%
		Relativistic electron velocity $v/c = 0.7$.		
60°	coordinate	+0.00%	-10.72%	-10.72%
60°	velocity	+40.01%	+15.03%	-21.88%
60°	both	+40.01%	+ 4.30%	-30.36%
"Experiment"		+40.03%	+0.00%	-28.59%

(Blatt and Weisskopf, 1952, pp. 785-789). In a magnetic field, it carries out a Larmor precession around the quantization axis (Goldstein, 1950, pp. 176-178). In order to study this effect, a "Larmor velocity component" was added to the above equations, and calculations were carried out using a polarization angle $\theta = 54.7°$. For Larmor angular velocities that are small in comparison to the spin angular velocity (0.1% or less), the effect of this additional Larmor velocity component is negligible. Thus the relativistic invariance of the spinning sphere model that we have demonstrated here should apply whether the "classical" (Chapter 11) or the "quantum mechanical" (Chapter 14) version is used.

One final relativistic test we can give this spinning sphere model is to spin the sphere at less than the relativistic limit (which is the limit where the equatorial speed reaches, or is infinitesimally below, the velocity of light, *c*). This lowered spin value alters the gyromagnetic ratio, and it also changes the relativistic transformation properties. To carry out this test, we took the example described in Tables 12.1 and 12.2 and applied it to the case where the equator is moving at half the velocity of light, $v = 0.5\,c$. This gave too low a spin value, which could be corrected by increasing the sphere radius. But it gave a calculated gyromagnetic ratio $g = 2.43$, which is too large, and which cannot be similarly corrected. Moreover, it gave the proper mass and spin transformation values, but it gave magnetic moment transformation values which are *(a)* correct at 0°, and *(b)* too large at 90° by a factor of almost two. Hence we have two important conclusions:

(1) the correct *spectroscopic properties* for this electron model are obtained only when the sphere is spinning at the *relativistic limit*;

(2) the correct *relativistic transformation properties* for this electron model are obtained only when the sphere is spinning at the *relativistic limit.*

Thus spin quantization is a necessary feature for this spatially extended model of the electron.

The main purpose of the studies described in the present chapter was to establish the fact that a spatially extended electron model possesses the proper Lorentz transformation properties. In the next chapter we examine another critical feature of this model: namely, the manner in which it leads to a two-component representation of the rotation group.

Notes

1 See Eqs. (11.15-11.19) in Chapter 11.

2 The relativistic sphere calculations described here, which used 40,000 or more mass units in the sphere, exceeded the available storage capacity on a CRAY-1 computer. They were carried out on a SUN SPARC station.

Chapter 13

Spatial Quantization and the Two-Component Rotation Group

One of the characteristic features of quantum mechanical systems is spatial quantization. Atomic orbitals with non-zero angular momentum values do not assume arbitrary orientations in an external magnetic field, but rather a series of quantized angles. This is a consequence of the symmetries of the orbital wave functions. Similarly, an electron in an atomic orbital does not have arbitrary spin angles, but just two – spin up and spin down with respect to the orbital quantization axis. Can we account for this electron behavior in terms of the relativistically-spinning-sphere electron model? It turns out that there is an interesting relationship which ties together the spin orientation of the electron, its electric quadrupole moment (or, more accurately, the vanishing of its electric quadrupole moment), and the conservation of orbital angular momentum. This is the topic of the present chapter. Historically, the two-valuedness of the electron spin was the clue that led directly to quantum electrodynamics – the quantum mechanical dynamics of the electron.

Electron spin is a concept that was deduced phenomenologically, as we discussed briefly in Chapter 10. It is of interest here to add in a few details about that discovery.[1] The need for the additional degree of freedom provided by the electron spin was first indicated by the widespread occurrence of doubling in the atomic spectral lines. But there were other unresolved problems. The most puzzling situation arose in connection with the observation of the *anomalous Zeeman effect* in the alkali metals. The alkali atoms each consist of an atomic "core" surrounded by a single outer valence electron, and they give rise to a complicated multiplet structure that could not be explained by any of the early theories, which were all based on the use of *integral* quantum numbers. A decisive advance was made by Alfred Landé, who deduced a formula that was based on *half-integral* quantum numbers, and which accurately reproduced the anomalous Zeeman splitting. Since electron orbitals all have integral quantum numbers, the half-integral component had to come from somewhere else. The first guess was that it must come from the atomic core. However, Pauli soon demonstrated that this could not be correct. Hence the half-integral component had to come from the outer valence electron in the alkali atom. The lowest half-integral value is 1/2, which, according to the usual vector rules, corresponds to a system with two degrees of freedom: + 1/2 and - 1/2. This led Pauli to postulate that the electron itself possesses this property. He described the situation as follows:[1]

(The anomalous Zeeman effect) "according to this point of view is due to a peculiar not classically describable two-valuedness of the quantum theoretical properties of the valency electron".

Soon afterwards, as we outlined in Chapter 10, first Kronig, and then independently Uhlenbeck and Goudsmit, took the logical next step and attributed this "two-valuedness" to the spin of the electron.

The hypothesis of electron spin actually involves three assumptions (van der Waerden, 1960, pp. 199-244):

(a) *the electron rotates*;
(b) *it has an angular momentum* $J_z = \pm 1/2\hbar$ *in a given direction*;
(c) *it has a magnetic moment* $\mu_z = (e/mc) \cdot J_z$.

Pauli was initially reluctant to accept *(a)* because of its "classical mechanical" character. He could have accepted *(b)* and *(c)* without accepting *(a)*, although it is not easy to visualize an angular momentum that does not correspond to a rotation. But Pauli also had reservations about *(b)* and *(c)*, because the calculated spin-orbit splitting was too large by a factor of two. However, this last objection was removed with the subsequent discovery of the "Thomas factor" (Thomas, 1926). Kronig, Uhlenbeck, and Goudsmit all considered *(a)*, *(b)*, and *(c)* to be inseparably linked together. The difficulty with a *classical mechanical* model is the following. If the electron appears as a two-valued object in a magnetic field, then a model that purports to represent the electron must also appear as a two-valued object in a magnetic field. That is, it must have a *quadrupole* type of interaction with the magnetic field or with the atom. But if we picture the electron as a spherically symmetric distribution of matter in which the ratio of the charge to the mass is everywhere the same, which is the usual *classical mechanical* picture that is invoked, then it has no electric quadrupole moment. Furthermore, as we showed in Eq. (11.10), this type of model yields the wrong gyromagnetic ratio for the electron. A spinning electrically charged spherical mass gives rise to a magnetic *dipole* moment, but there is no obvious reason why this dipole moment should behave in a quadrupole-like manner in an external magnetic field. Thus it seems as if a classical representation of the electron is not possible. And, indeed, this is the prevailing view of most physicists today. We repeat here the quotation by van der Waerden (1960, pp. 215-216) that was given in Chapter 10:

"The spin (of the electron) cannot be described by a classical kinematic model, for such a model can never lead to a two-valued representation of the rotation group."

The electron thus presents us with a real puzzle. In order to behave in a two-valued manner in an atom, the electron must exhibit a quadrupole interaction of some type. But we know empirically that the electron does not have an observable *electric* quadrupole moment (or it would appear in experiments with a finite size), and its *magnetic* moment is a dipole rather than a quadrupole. Also, from a theoretical point of view, models of the electron, as noted above, do not naturally give rise to quadrupole interactions. The *two-component behavior* of the electron in atomic spectra, which mandates a quadrupole effect, and which was the *sine qua non* for introducing electron spin in the first place, seems to rule out any kind of a classical representation to account for the spin.

The answer to this puzzle is another puzzle: *we need an electron quadrupole moment that is not a quadrupole moment.* Or, more correctly, we need an electron quadrupole moment that is not an *observable* quadrupole moment. Rather surprisingly, it is possible to achieve this result. In fact, we have already achieved it. This is the property of the *quantum current loop* that we described in Chapter 7. Let us invoke Ampere's hypothesis and attribute the magnetic field of the electron to a current loop. This current loop gives rise to a sequence of even-numbered electric moments: monopole V_0, quadrupole V_2, ... (Eq. 7.4). Now, according to the vector formalism of quantum mechanics, the electron spin axis, which is also the axis of the current loop, is not directed along the z-axis of quantization, but instead is inclined at an angle

$$\theta_{QM} = \pm \arccos \frac{1/2}{\sqrt{1/2(1/2+1)}} = \pm 54.7° . \tag{13.1}$$

As we demonstrated in Chapter 7, the electric quadrupole moment V_2 vanishes at these angles. Specifically, it vanishes identically along the z axis, and it vanishes along the x and y axes when averaged over a cycle of precessional motion. Hence, if we place a spinning electron in a magnetic field, there are two angles at which its electric quadrupole moment vanishes, namely at $\pm 54.7°$. This result logically gives rise to the two-valued behavior of the electron: *the electron always orients itself in a quantizing magnetic field so as to have a vanishing electric quadrupole moment.*

We have now *accounted-for* the spatial quantization of the electron, but we have not yet *explained* it. We have an effect without a cause. We are in somewhat the same position as was Landé when he derived his empirical expression for the splitting of the anomalous Zeeman levels in the alkali atoms. The Landé equation worked so well that it was indisputably correct, but the physical underpinning had not yet been provided. We cannot provide this underpinning as concretely as did Pauli, Kronig, Uhlenbeck, Goudsmit, and Thomas, but we can at least make the result plausible. To do this, we must move on to the subsequent development of wave mechanics by de Broglie, Schrödinger, Heisenberg, and Dirac. These workers demonstrated that when an electron moves, it generates a wave which accompanies and in some manner influences the motion of the electron. In particular, the symmetry requirements of the wave must be met in order to have viable particle trajectories. This is perhaps most readily seen in the case of bound atomic orbitals. Bohr showed that electrons exist in quantized orbits in atoms, and he supplied the key discovery that *these orbitals correspond to quantized values of the orbital angular momentum.* However, he had no physical model to explain why this was so. When de Broglie deduced his electron wave equation, $\lambda = h/p$, he immediately applied it to the atom and demonstrated the fact that the quantized orbits are the ones in which the wave and the electron remain in phase. Schrödinger extended de Broglie's wave picture, which was based on plane waves, by modifying the waves so as to apply directly to the case of an electron bound in a potential well. The "Schrödinger wave function" ψ that he obtained contains the full information about the electron orbitals. The spatial derivatives of the ψ wave give the momentum components of the electron, and the time derivative gives its energy. In fact, the electron itself no longer appears in the Schrödinger formulation: the ψ wave is all that is necessary for the solution of the problem. The product $\psi^*\psi$ is a probability density that gives the probability of finding the electron at any location inside the wave envelope. Several constraints are imposed on the ψ wave: energy, momentum, and angular momentum must be conserved; the wave and its derivatives must be continuous; the wave must be single-valued, which means that it closes on itself to form a uniform standing wave pattern for the orbital; and the probability density $\psi^*\psi$ must be equal to unity when integrated over all of space, since it represents a single electron. It turns out that these conditions, together with the form of the potential that binds the atom, are sufficient to completely specify the wave. The result of all of these conditions is that only electrons with suitable energy, momentum, and angular momentum values can form stable bound-state orbitals inside an atom.

The hydrogen atom is the simplest example of a bound atomic system. We can use time-independent ψ waves to illustrate the nature of the stable atomic orbitals. The Schrödinger equation for hydrogen features a Coulomb potential, $V(r)$, that binds the electron to the proton, with the Coulomb force depending only on the radial coordinate r, so that it is spherically symmetric. As shown for example in Schiff (1955, pp. 69-73), the ψ wave for a spherically symmetric potential separates into radial and angular coordinates:

$$\psi_{nlm} = R_{nl}(r)\, Y_{lm}(\theta,\phi) \quad , \tag{13.2}$$

where (r,θ,ϕ) are the usual spherical coordinates. The hydrogen spectral quantum numbers n,l,m of the electron are related to its total energy, orbital angular momentum, and projection of the orbital angular momentum along the quantization axis, respectively. This is a rather schematic presentation of the Schrödinger wave function for hydrogen, but it serves to bring out the point we now wish to make.

Consider an $l = 1$ P-wave orbit in hydrogen. According to the rules of quantum mechanics, if we apply an external magnetic field, then this P-wave electron orbit must assume one of three quantized spatial orientations, which are specified by the quantum numbers $m = +1$, 0, or -1. Now, the electron in the P-wave orbit generates a magnetic field, and this magnetic field interacts with the applied external magnetic field. But the magnetic interaction *per se* is *not* what is producing these three spatial orientations of the orbit. These three spatial orientations exist because they are the three ways that an electron with this energy and angular momentum can orbit so as to produce a standing ψ wave with the required symmetry properties.

It is the symmetry requirements of the ψ wave that lead to the spatial quantization. This is a purely quantum phenomenon that cannot be accounted-for classically.

The point of this discussion is that we can see the same situation – that is, the same wave function symmetry effect – occurring in the case of the electron spin. If the electron has no electric quadrupole moment, then its Coulomb interaction with the nucleus is spherically symmetric, and we obtain the solutions that have just been described. But if it possesses an electric quadrupole moment, then the Coulomb interaction is no longer spherically symmetric, and the electron moves in a slightly perturbed orbit which does not have the requisite ψ-wave symmetry properties. Thus the electron, in order to maintain a stable orbit, must minimize the effect of the quadrupole perturbation. This it can do by assuming one of two spin angles with respect to the quantization axis, $\theta_{QM} = \pm 54.7°$, which it does. Hence it appears as a two-valued member of the three-dimensional rotation group.

We can supply a physical basis for this result. Bohr first showed that the quantization of atomic orbitals corresponds to the quantization of orbital angular momentum, which must therefore be a conserved quantity. In the Schrödinger equation, the separation of the wave function into separate radial and orbital wave functions can *only* be carried out for the case of a central potential. This separation has real physical implications: *it corresponds to the conservation of angular momentum.* Schiff (1955, p. 75) describes the situation as follows:

"It should be noted that the wave equation cannot in general be separated in this way and angular-momentum eigenfunctions obtained if the potential energy $V(r)$ is not spherically symmetric. This corresponds to the classical result that the angular momentum is a constant of motion only for a central field of force (which is describable by a spherically symmetric potential)."

Thus, to the extent that the electron has a non-central electric quadrupole interaction with the nucleus, angular momentum is not conserved, and a stable ψ wave cannot be formed. Hence the stable orbits are the orbits in which the electric quadrupole effect cancels out.[2] This occurs at the quantum-mechanical angles $\pm\, \theta_{QM}$.

It is instructive to consider the electron quadrupole effect in more detail. The electron model that we arrive at in Chapters 11 and 14 consists of a spinning mechanical mass with a point charge on its equator. The rotational frequency of the electron is greater than 10^{20} Hz, as compared to a typical orbital frequency in the atom of about 10^{15} Hz.[3] Thus the point charge on the electron functions effectively as a current loop with respect to its effect on the atomic orbitals. The magnetic moment of this current loop interacts with the magnetic field of the orbital motion and causes a Larmor precession. Typical Larmor frequencies in the atom are roughly 10^{10} Hz,[3] and hence are much smaller than orbital frequencies. Thus the current loop on the electron has essentially a fixed orientation in space as it travels a cycle of revolution around the atomic center. As a result, the precessional motion of the current loop, which is necessary in order to average out the electric quadrupole potential along the x and y axes, actually comes in this case from the orbital rather than the Larmor rotation. At the quantization angles θ_{QM}, this averaging causes the electron to operate with essentially the same central potential V_0 (Eq. 7.5) that it would have if the electron were a point. Hence the electron maintains the same average radius r with respect to the position of the nucleus. At angles other than θ_{QM}, the quadrupole effect does not vanish, and the electron is forced to a slightly different average radius, and hence a slightly different value for the angular momentum and somewhat different closure properties. These perturbations evidently have an important effect on the symmetry of the ψ wave, and they lead to the result that stable orbits feature spin orientation angles of $\pm 54.7°$ with respect to the axis of rotation.

In this discussion, we have been implicitly taking the Bohr concept of a localized electron revolving in an orbit around an atomic nucleus quite literally. This might seem to be at variance with the Schrödinger formalism that is used to reproduce atomic energy levels, because the Schrödinger wave functions do not delineate any specific rotational motion of the electron. Physicists who emphasize the wave aspects of the electron characterize it as having an *amoeba-like character* (Jaynes, 1991, p. 9), in which a non-localized bound-state electron simply surrounds and envelops the atomic nucleus. However, physicists who argue the particle point-of-view consider the atomic orbitals to be real physical entities. As Hestenes (1991, p. 32) points out,[4]

> " ... the time dilatation in the decay of μ^- particles captured in atomic s-states indicates that they *really are moving with the Bohr velocity* in those states (Silverman, 1982). So must electrons move also."

Thus the Bohr configuration-space linear velocities seem to be as significant as the Bohr momentum-space angular velocities. These velocities are indicative of localized electrons that move in localized trajectories.

The effects we have just considered involve motions within an atom. What about the motion of the entire atom itself? In the Stern-Gerlach experiment, a spin 1/2 atom, which consists of a single outer spin 1/2 electron and a spin 0 "atomic core", is observed to split into two spatial components when passed through an inhomogeneous magnetic field. This splitting is attributable to the two spatial orientations of the electron spin vector. But whereas the electron we discussed above was quantized with respect to the atomic orbitals inside an atom, it is now quantized with respect to the external magnetic field. What is the effect of the electric quadrupole moment of the electron on this external magnetic field? The key point here seems to be the fact that in the reference frame of the atom, the Stern-Gerlach magnetic field contains both magnetic and electric components (Jackson, 1962, p. 380). The interaction of the electric quadrupole moment of the electron with the external *electric* field causes wave function perturbations, which are minimized when it is oriented at the quantization angles θ_{QM}. The *magnetic* interactions then function so as to split the atomic trajectories into components that correspond to these two spin orientations.

The discovery of the two-component behavior of the electron, with its *up* and *down* spin orientations, cleared up the difficulties with respect to atomic spectra. But it accomplished a lot more than just that. It led directly to the theory of quantum electrodynamics, and to the discovery of the positron. Pauli derived a set of three spin matrices that constitute a two-valued representation of the three-dimensional rotation group:

$$\sigma_X = \begin{pmatrix} 0 & 1 \\ 1 & 0 \end{pmatrix}; \quad \sigma_Y = \begin{pmatrix} 0 & -i \\ i & 0 \end{pmatrix}; \quad \sigma_Z = \begin{pmatrix} 1 & 0 \\ 0 & -1 \end{pmatrix}, \tag{13.3}$$

where the matrix elements are in units of $1/2\,\hbar$. These matrices give the projection of the electron spin along the z axis as

$$J_Z = \pm\frac{1}{2}\hbar , \tag{13.4}$$

and they give vanishing expectation values for the spin along the x and y axes. They also show that the total electron spin angular momentum is

$$J = \sqrt{\sigma_X^2 + \sigma_Y^2 + \sigma_Z^2} \cdot \frac{1}{2}\hbar = \frac{\sqrt{3}}{2}\hbar . \tag{13.5}$$

These *Pauli spin matrices* thus reproduce the quantum mechanical features of the electron spin. However, they are not relativistic. The difficulty is that we require *four* anticommuting matrices in order to construct a Lorentz-invariant four-vector, and no other 2 x 2 matrices can be found that anticommute with the three Pauli matrices. The solution to this difficulty was found by Dirac (1928). Seeking a relativistic form of the electron wave equation, he realized that he must use 4 x 4 rather than 2 x 2 matrices. The Dirac spinors that correspond to the 4 x 4 Dirac matrices turned out to represent not only the spin orientations (up or down) of the electron, but also those of its antiparticle – the *positron*. Thus they led to the prediction of the positron, which was discovered by Anderson (1932). The Dirac equation for the electron is the relativistic counterpart of the non-relativistic Schrödinger equation, and it laid the foundation for quantum electrodynamics, which is arguably the most accurate formalism ever developed in physics. This chain of events was set in motion by the simple observation that the electron must have a two-valued representation in atomic orbitals.

In contrast to the discussions of the previous chapters, which feature definite mathematical results, the discussion in this chapter has been largely in the form of plausibility arguments. What we have been seeking to establish here is the fact that a "classical mechanical" model of the electron can give rise to quadrupole effects, and hence to a "two-valuedness", which is what we phenomenologically require for the electron spin. But this quest led us to a conceptual dilemma: although we need a quadrupole effect in order to obtain a two-component electron spin, the existence of this required quadrupole interaction (and hence of a corresponding non-central force) is inconsistent with the observed point-like nature of the electron, and also with the conservation of angular momentum in atomic orbitals. The way out of this dilemma is provided rather miraculously by the strange behavior of the electric quadrupole moment of the electron, which vanishes at the quantization angle at which we must try to measure it. *The required quadrupole quantization effect is provided by the **vanishing** of the quadrupole moment.*

The spin systematics of the electron comes out of the rules of quantum mechanics, which were deduced without any specific electron model in mind. The prevalent view is that we should not – indeed, we cannot – provide a physical model, and especially a "classical" physical model, that purports to account for these quantum mechanical rules. In general, it does not seem to be necessary to go into the details of such a model. For instance, Schulman, in deriving a path integral formulation for spinning particles, comments as follows (Schulman, 1968, p. 1559):

> "... we consider this angular momentum to be the angular momentum of *something*; something is spinning. This, however, is just another way of saying that there is an internal variable."

By expanding his calculational space to include this new degree of freedom, Schulman arrives at a quantized formalism which includes, but is not limited to, spin 1/2 particles. The electron model that we have described here should in a sense be regarded as an extension of this type of formal approach. The model must be consistent with the constraints imposed by these generalized formalisms, and if it is, then it has justified its right to co-existence.

Notes

[1] The discussion at the beginning of this chapter about the discovery of electron spin is taken mainly from Pais (1986, Chapter 13), which contains references to the early literature.

[2] We also know from the workings of the Pauli Exclusion Principle that the stable atomic orbits each contain only one electron. Two orbiting electrons whose total (spatial plus spin) wave functions are orthogonal do not, on the average, interfere with one another; that is, they do not, on the average, exchange energy or momentum.

[3] See Cagnac and Pebay-Peyroula (1975, p. 200).

[4] The (Silverman, 1982) reference in this quotation was changed to our bibliographic format.

Chapter 14

A Quantum-Mechanical Model of the Electron

This is the chapter in which we finally put together all of the spectroscopic information about the electron. More exactly, this is the chapter in which we incorporate these spectroscopic features into the electron model. Let us proceed in a stepwise fashion, adding each feature in accordance with the experimental and theoretical demands on the model.

We start with a uniform sphere of *mechanical matter*. Its rest mass is

$$m_0 = \frac{2}{3}m \cong 0.341 \text{ MeV/c}^2 \quad , \tag{14.1}$$

where m is the measured mass of the electron (Table 9.1). We then spin the sphere until it reaches the relativistic limit, with the equator moving at, or infinitesimally below, the velocity of light c. The *mass* of the spinning sphere is increased relativistically by a factor of 3/2 (Eq. 10.5), so that the spinning mass m_S of the electron in the laboratory frame of reference is

$$m_S = m \cong 0.511 \text{ MeV/c}^2 \quad . \tag{14.2}$$

The relativistic increase of each mass element in the sphere (Eq. 10.1) can be pictured as arising either from special relativity (via the instantaneous velocity of each element) or from general relativity (via the effective gravitational potential associated with a rotating coordinate system). The spherical envelope that represents the *shape* of the sphere is not altered relativistically, since the rotational motion is at right angles to the sphere radius, which is therefore an invariant. However, the measured *volume* of the sphere is increased by a factor of 3/2, due to the perceived non-Euclidean geometry inside the rotating sphere (Chapters 10 and 11). Thus the *density* of the mechanical matter remains unchanged, which indicates that the centrifugal effects of the rotation are incorporated in the relativistic increase of the mass. The continuous nature of the integration over the mass elements within the sphere (Eq. 10.3) shows that the microscopic mechanical mass of the electron forms a *continuum*, which is a state of matter that has no macroscopic counterpart (Chapter 15). The freedom from distortional effects suggests that this mechanical mass fits the definition in classical mechanics of a rigid body (Chapter 15).

The relativistic moment of inertia of the spinning sphere about the axis of rotation is given by Eq. (10.6), and is equal to

$$I = \frac{3}{4} m_0 R^2 = \frac{1}{2} m R^2 , \tag{14.3}$$

where R is the radius of the sphere. Setting R equal to the Compton radius,

$$R_C = \frac{\hbar}{mc} , \tag{14.4}$$

we obtain

$$I = \frac{1}{2} \frac{\hbar^2}{mc^2} . \tag{14.5}$$

The angular momentum of the spinning sphere is equal to

$$J = I \omega , \tag{14.6}$$

so that if we set ω equal to its relativistically limiting value,

$$\omega = \frac{c}{R_C} = \frac{mc^2}{\hbar} , \tag{14.7}$$

we obtain

$$J_{obs} = \frac{1}{2} \hbar \tag{14.8}$$

as a directly calculated quantity. Hence it is clear that a straightforward classical model exists which can reproduce the observed spin of the electron.

However, there is an immediate complication. As we discussed in Chapter 13, the quantum mechanical formalism of angular momentum vectors, which is embodied in the Pauli spin matrices, shows that the *total* spin angular momentum of the electron is

$$J = \sqrt{\frac{1}{2}\left(\frac{1}{2}+1\right)} \hbar = \frac{\sqrt{3}}{2} \hbar , \tag{14.9}$$

and that the value of 1/2 $\hbar$ shown in Eq. (14.8) is the projection of the spin along the z-axis of quantization (the observed value). To incorporate this result into the electron model, we first increase its radius by a factor of $\sqrt{3}$, thus obtaining the "quantum-mechanical Compton radius"

$$R_{QMC} = \sqrt{3} \frac{\hbar}{mc} . \tag{14.10}$$

Since the relativistically spinning sphere model gives the general result that

$$J = \frac{1}{2} mRc , \tag{14.11}$$

we see that increasing the radius by a factor of $\sqrt{3}$ increases the spin angular momentum by this same factor. Then, in order to correctly project this spin angular momentum onto the quantization axis, we rotate the spin axis away from the z axis to one of the quantum-mechanically prescribed angles,

$$\theta_{QM} = \pm \arccos \frac{1}{\sqrt{3}} \cong \pm 54.7° \quad . \tag{14.12}$$

In this manner we have, at least phenomenologically, reproduced the quantum mechanical properties of the electron spin. But an obvious difficulty with this purely mechanical model is that, as it stands, the spinning sphere has no quadrupole moment with which to account for the two quantization angles shown in Eq. (14.12). We can remedy this defect by adding in electromagnetic effects. These of course are also required in order to reproduce the charge and magnetic moment of the electron.

With respect to quantum-mechanical projection factors, the magnetic properties of the electron necessarily echo its angular momentum properties. That is, the observed magnetic moment of the electron is

$$\mu_Z = \frac{e\hbar}{2mc} \quad , \tag{14.13}$$

and its total magnetic moment is $\sqrt{3}$ larger. In order to reproduce these values, we place a charge e on the *equator* of the spinning sphere. This rotating charge constitutes a current loop. If the charge e is moving at the velocity c, then the magnetic moment that corresponds to this current loop is, from Eq. (11.17),

$$\mu = \frac{eR_{QMC}}{2} = \sqrt{3}\,\frac{e\hbar}{2mc} \quad . \tag{14.14}$$

Thus this model gives the correct quantum mechanical behavior for the electron magnetic moment. The gyromagnetic ratio of the electron,

$$g = \frac{\mu}{J} \cdot \frac{2mc}{e} = 2 \quad , \tag{14.15}$$

follows directly from Eqs. (14.9) and (14.14). Hence we have now reproduced the salient spectroscopic features of the electron as they were known *circa* 1935. We have also disproved the long-held belief that no classical model can reproduce the gyromagnetic ratio of the electron.[1]

In addition to providing the charge and the magnetic moment of the electron, the other function provided by the equatorial charge e is to supply the quadrupole-type of interaction that is required in order to account for the quantization angles $\theta_{QM} = \pm 54.7°$ of Eq. (14.12). As we discussed in Chapters 7 and 13, the equatorial current loop on the electron functions as a *quantum current loop*. That is, its electric quadrupole moment – the term V_2 in Eqs. (7.4) and (7.6) – vanishes at the angles $\pm\,\theta_{QM}$. A non-vanishing electric quadrupole moment constitutes a non-central force, which leads to the non-conservation of angular momentum (Chapter 13). Since angular momentum is one of the conserved quantum numbers required for the construction of a stable electron orbital, an electron occupying such an orbital is forced to minimize its electric quadrupole moment. This it accomplishes by having the spin axis oriented at an angle θ_{QM}. Thus we have a viable mechanism that enables us to use this *classical* electron model as the basis for explaining the *quantum-mechanical* spin behavior of the electron.

The current loop on the electron could in principle be in the form of either a single point charge *e*, or else a distributed ring of charge. However, in order to obtain the above results, we require that the electrostatic self-energy W_E of this charge be equal to zero, since a non-zero charge would contribute to the inertial properties of the spinning sphere and thus alter the angular momentum equations. Furthermore, the very small size of this charge,

$$R_E \; < \; 10^{-16}\text{cm} \quad , \tag{7.18}$$

mandates that its self-energy, if viewed classically, must be identically zero (see questions (7.13) and (7.13a) and Eq. (7.20) in Chapter 7). Also, the fact that this charge *e* is moving at the velocity *c* suggests that it does not contribute to the energy, since any finite contribution would be expected to diverge mathematically. These considerations indicate that the charge *e* is in the form of a single non-interacting point, since the various segments of a charged ring would logically interact with one another. The electron scattering calculations described in Chapter 16 also suggest that the charge *e* is a point.

Although there are several reasons for thinking that the electrostatic self-energy W_E of the electron is zero, these reasons do not apply to the magnetic self-energy W_H. The magnetic field of the electron, as Rasetti and Fermi (1926) first noted, must extend over a region of space which is much larger than the size of the electric charge. In the present model, it extends over a region the size of the *quantum current loop*, whose radius is

$$R_{QMC} \; \cong \; 6.7 \times 10^{-11}\ \text{cm} \quad . \tag{14.16}$$

Extending the classical estimate of Fermi and Rasetti, we calculated the self-energy of this magnetic field in Chapter 8 as

$$W_H \; \geq \; 359\ \text{eV} \quad , \tag{8.16$'$}$$

so that

$$\frac{W_H}{mc^2} \; \geq \; 0.07\% \quad . \tag{14.17}$$

This is a crucial result, because it forces us to modify the electron model. In deriving the spin angular momentum (Eqs. 14.3 - 14.8), we assumed that all of the observed mass *m* (Eq. 14.2) contributes to the rotational motion. However, the magnetic field of the electron is *irrotational.* Since the magnetic self-energy constitutes at least 0.07% of the total mass of the electron, we must reduce the *calculated* spin angular momentum of the model by this amount, so that the model now gives a slightly incorrect (low) value for the spin. The calculated value of the magnetic moment (Eq. 14.14) depends only on the radius of the sphere, which we for the moment hold constant. Thus the magnetic moment is unaffected by this change, so that the calculated value for the gyromagnetic ratio, which is the ratio of these two quantities, is altered (in the correct direction), and it becomes

$$g \; \geq \; 2 \times 1.0007 \quad . \tag{14.18}$$

This is a calculation (Mac Gregor, 1989) that could have been applied to the electron many years ago, and which may have conceivably led to the prediction of this g-value anomaly. Historically, in the year 1938 discrepancies began to show up in some spectral *g*-values. These were clarified experimentally by the work of Rabi and coworkers (Nafe, 1947), and were then interpreted by Breit (1947) as arising from an electron magnetic moment that is about 0.1% larger than the value given in Eq. (14.13) (see Chapters 3 and 8). Calculations by Schwinger (1948) established the magnitude of this effect as

$$\frac{\alpha}{2\pi} \cong 0.116\% \quad . \tag{14.19}$$

A series of increasingly precise measurements of the gyromagnetic ratio of the electron have confirmed this result, finally giving (Table 9.1)

$$g = 2 \times 1.001159652193 \quad . \tag{14.20}$$

We have quoted this result in full detail to illustrate the incredible precision of not only the experimental measurements, but also the QED theory, which is in complete agreement with experiment (Chapter 9). The essential point here is that the experimental g ratio shown in Eq. (14.20) is very close to the rough estimate given in Eq. (14.18), which corresponds to a *lower bound* on the magnetic self-energy of the electron. This suggests that the magnetic self-energy of the electron is accurately specified by Eq. (4.19) to be

$$W_H \cong \frac{\alpha}{2\pi} \cdot mc^2 = 593 \text{ eV} \quad , \tag{14.21}$$

and that this correction should be applied to the electron model. This requires two changes in the model. The spinning mechanical mass becomes

$$m_S \equiv m \cdot \left(1 - \frac{\alpha}{2\pi}\right) = 0.51041 \text{ MeV}/c^2 \quad , \tag{14.22}$$

and the sphere radius becomes

$$R^{\alpha}_{QMC} \equiv \sqrt{3}\frac{\hbar}{mc} \cdot \left(1 + \frac{\alpha}{2\pi}\right) = 6.6962 \times 10^{-11} \text{ cm} \quad . \tag{14.23}$$

With these changes, the calculated spin is maintained at its proper value (Eqs. 14.8 and 14.9), and the calculated magnetic moment is increased so as to reflect the effect of the anomalous magnetic moment (Eq. 8.10). The magnetic self-energy correction of 593 eV that we have adopted here is reasonably close to the lower-bound estimate of 359 eV shown in Eq. (8.16′).

The quantum-mechanical electron model that we finally obtain is described in Table 14.1, and is displayed graphically in Figure 14.1. As shown in Table 14.1 and Figure 14.1, this electron model is quite simple, and it reproduces the standard features of the electron. Furthermore, it accomplishes this with essentially no arbitrary parameters. In order to substantiate this remark, we need to examine the degrees of freedom in the model. If we ignore the small correction for the magnetic self-energy, then the only free parameter in the model is the radius R, and the model reproduces two quantities – the spin and the Dirac magnetic moment of the electron – that both depend on R. Hence it is an overconstrained model, and the fact that a single value for R fits both of these quantities is a non-trivial result. The relativistic aspects of the spinning sphere are of crucial importance in obtaining this result. Without these relativistic effects, the model does not work, which is the basis for the frequently made assertion that such a model is not possible.[1] If we add in the magnetic self-energy W_H as another adjustable parameter, then the model has two degrees of

TABLE 14.1 THE QUANTUM-MECHANICAL MODEL OF THE ELECTRON

Non-rotating rest frame properties	
$m_0 = m \cdot (1-\alpha/2\pi) \cdot 2/3$	(m_0 = mechanical mass)
$R = \sqrt{3} \cdot \hbar/mc \cdot (1+\alpha/2\pi)$	(m = experimental mass)
$W_E = 0$	(W_E = electrostatic self-energy)
e = equatorial unit point charge	
Calculated rotating inertial properties	
$m_S = m \cdot (1-\alpha/2\pi)$	(m_S = spinning mechanical mass)
$W_H = mc^2 \cdot (\alpha/2\pi)$	(W_H = magnetic self-energy)
$I = 1/2 \cdot m_S R^2$	($\omega = c/R$) (relativistic limit)
Calculated spectroscopic quantities	
$J = \sqrt{3}/2 \cdot \hbar$	(vanishing electric dipole moment)
$\mu = \sqrt{3} \cdot e\hbar/2mc \cdot (1+\alpha/2\pi)$	(non-vanishing electric quadrupole moment)
Spectroscopic quantities at the quantization angle θ_{QM}	
$J_Z = 1/2 \cdot \hbar$	($\theta_{QM} = \pm 54.7°$)
$\mu_Z = e\hbar/2mc \cdot (1+\alpha/2\pi)$	(vanishing electric quadrupole moment)

freedom, R and W_H, and it reproduces three quantities – the spin, the Dirac magnetic moment, and the anomalous magnetic moment of the electron. Hence it is still overconstrained. One might argue that the Dirac magnetic moment and the anomalous magnetic moment together represent just one quantity, but it can then also be argued that the magnetic self-energy is to some extent calculable, and hence not really a free parameter. Also, the magnetic self-energy is subjected to the constraint that the observed mass of the electron must be reproduced.

One conclusion we can draw from the present studies is that this electron model actually does exist. This might seem like a tautological statement, but it is not. A massive body of literature about the electron has been generated over the past century, but none of it, except for work by the present author, contains any reference to such a model. The mathematical equations that are given here have been fully refereed, and they are correct. The model does what it is supposed to do. Whether this model has any contact with reality is of course a completely different question. However, the usefulness of the model in a sense transcends this question. An electron model that gives correct spectroscopic answers ought to become a part of the general lore of the particle physicist.

Another conclusion we obtain – a conclusion as to the size of the electron – is one that impacts directly on the physical reality of this electron model. If the only free parameter

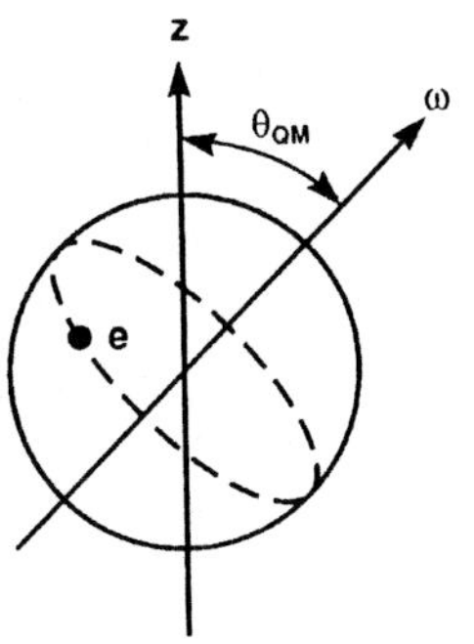

Fig. 14.1. The quantum-mechanical electron model. It consists of a mechanical sphere of matter with a point charge *e* on the equator. The rest mass and radius of the sphere are $m_0 = 340{,}270$ eV/c^2 and R = 6.6962 x 10^{-11} cm, as defined in Table 14.1. When spun at the relativistic limit, this sphere reproduces the *total* spin and magnetic moment of the electron, and hence its gyromagnetic ratio, correctly to first order in α. When the magnetic self-energy W_H from Table 14.1 is added in, it also reproduces the observed mass of the electron. And when the relativistically spinning sphere is tipped at an angle θ_{QM} with respect to the quantization axis, it reproduces the *observed* spin and magnetic moment of the electron, and it has an electric quadrupole moment that averages out to zero over a cycle of precessional motion. Furthermore, the mass, spin, and magnetic moment of this electron model transform correctly under Lorentz transformations, as was demonstrated in Chapter 12. Its scattering properties, which are point-like in nature except possibly in a narrow keV energy window, are calculated in Chapter 16 and discussed in Chapter 17.

in this model is the electron radius R, and if the model itself is overconstrained, then the value that it yields for R must be unambiguous. And it is. The electron, according to this viewpoint, has a radius

$$R^{\alpha}_{QMC} \equiv 6.70 \times 10^{-11}\ \text{cm} \quad .$$

This is the only radius that gives a fit to the experimental spectroscopic data on the electron. But this is an absolutely enormous radius. Almost every physicist living today believes that the electron actually has a radius

$$R_{\text{electron}} < 10^{-16}\ \text{cm} \quad .$$

These two values differ by a factor of a million! How can an electron that seemingly interacts in all respects as a point be as large as the present model would have us believe? This is the topic that we address in the final chapters of this monograph. The burden of proof is clearly on the author. When we examine this situation computationally, some very interesting results emerge. This large electron does scatter in a point-like manner in the energy regions where it has been carefully measured, but it may exhibit finite-size effects in one rather difficult-to-reach intermediate energy region – the region of kilovolt bombarding energies – which is a region that has not been thoroughly investigated.

The concept of a large electron raises many problems, but the alternative concept of a point electron is not without its own difficulties. By postulating a point-like nature for the electron, physicists are sweeping all of the spectroscopy we know about the electron under the rug, and they are sweeping the classical physics of the microworld under the same rug.

Notes

[1] See Note 4 in Chapter 2.

Part IV

The Mott Channeling of Finite-Size Electrons

"... surprises in fields which had seemed well understood have frequently furthered progress in physics."

Joachim Kessler[1]

Chapter 15

Mechanical Mass: A New State of Matter

Broadly speaking, we have two kinds of information about the electron: its *spectroscopic* features – which are the properties of an isolated particle; and its *dynamical* features – which are the ways in which it interacts with other particles. We have now completed our examination of the spectroscopic aspects of the electron. These can be accurately reproduced (to first order in α) by a large (Compton-sized) relativistically spinning sphere of mechanical matter that contains an embedded equatorial point charge (Figure 14.1). The question which now arises is whether this electron model, which was derived mainly from spectroscopic considerations, can also account for the dynamical aspects of the electron – most importantly, its point-like scattering. This is the topic that occupies the remainder of the book. A definitive dynamical test is provided by Mott scattering off atomic nuclei. Computationally, we discover that this large electron model does lead to point-like Mott scattering in most energy regions, but not in the keV region, where the scattering impact parameters are comparable to the electron Compton radius. In this keV energy region, we seem to have direct manifestations of electron finite-size effects, both theoretically (Chapter 16) and possibly also experimentally (Chapter 17). The computer calculations, which are crucial to this analysis, cannot be carried out without a completely specified electron model. This requires that we delineate the dynamical properties of the mechanical mass: in particular, the manner in which it is mechanically coupled to the charge on the electron. Since these properties are *a priori* unknown, we must assemble all of the available information that has a bearing on this topic. This is the task of the present chapter.

The nature of the mass of the electron remains one of the unsolved problems in present-day elementary particle physics. This problem was summarized very succinctly by Pais in the quotation given in Chapter 1, to which we also referred in Chapter 9. As he stated it (Pais, 1982, p. 159) , we know from a number of arguments that

"the mass of the electron is certainly not purely electromagnetic in nature",

but we have no idea as to what the non-electromagnetic component of the electron is:

"... we still do not know what causes the electron to weigh".

If the electron is Compton-sized, then gravitational effects are miniscule, so this unknown non-electromagnetic mass is, by default, *mechanical.* One of the virtues of studying the spectroscopy of the electron in detail, as we have done in the preceding chapters, is that it provides at least some basis for investigating the possible features of this mechanical mass. The mass of the electron is of course directly related to its energy content, and in the present discussion we will tend to use the words *mass* and *energy* interchangeably. Let us see what we can conclude about the properties of mechanical masses. We first review the various possible contributors to the energy-content of the electron.

In Chapter 8 we wrote down four possible mass or energy components in the electron:

gravitational (W_G)*; electrostatic* (W_E)*; magnetic* (W_H)*; mechanical* (W_M)*.*

A *gravitational* mass component is a mass that interacts with itself via gravitational forces. In the case of a mass that is on the order of the electron mass, 9 x 10^{-28} gm, gravitational self-interactions are negligible unless we compress it down to a size of roughly 10^{-30} cm. Since this size is much smaller than any dimensions that we ascribe to the electron in the present studies, we will assume that the gravitational mass component W_G is completely negligible.

Now consider the electromagnetic mass components, W_E and W_H. The earliest electron models, which were formulated before the discovery of electron spin, were assumed to be purely *electrostatic*, as we discussed in Chapter 3. However, the stability problem of a purely electrostatic mass was never solved. Later, after the discovery of electron spin, a *magnetic* component in the electron was also recognized. But this still did not solve the stability problem. *No purely electromagnetic electron model is stable.* Some kind of non-electromagnetic "glue" – a Poincaré force – is required in order to hold the electron together. In the electron model that we have set forth in Chapters 11 and 14, the electric charge e is in the form of a single point, and it has a radius R_E that is known experimentally to be less than 10^{-16} cm (Eq. 7.18). Now, if this point charge were to interact with itself classically (Eqs. 7.19 and 7.20), its electrostatic self-energy W_E would be much larger than the total mass of the electron. Hence it is clear that the charge e does not interact in this manner. This solves part of the stability problem, in that the charge e (whatever it is) evidently holds itself together. But this leaves us with no calculational basis for assigning a value to W_E. It could be anywhere between zero and 0.511 MeV. However, the electron model itself comes to our rescue here. In this model, the charge e is perched on the equator of the spinning sphere, which is moving at the velocity c. Thus we must have $W_E = 0$ in order to keep the electron mass finite. Furthermore, the electron spin angular momentum is correctly calculated (Eqs. 10.6 - 10.11) with the assumption that W_E does not contribute. Hence we have several reasons for believing that the electrostatic self-energy W_E is equal to zero. The magnetic self-energy W_H, by way of contrast, is non-zero in the present model, but it amounts to only 0.1% of the total electron mass (Eq. 8.15). Since W_G, W_E, and W_H together comprise only 0.1% of the observed mass of the electron, we require the *non-electromagnetic* mass component W_M in order to account for other 99.9% of the mass. Furthermore, the attractive gravitational forces in the electron are negligible, and the electromagnetic forces are unstable, so that we require this non-electromagnetic mass component in order to hold the electron together. This mechanical mass evidently provides its own stability, and it holds the electric charge e in place.

We see from the above summary that *stability* and *energy* considerations both lead to the conclusion that the electron must have a non-gravitational, non-electromagnetic mass component. But what is this mass, and what are its properties? Conceptually, we start with the electron, and then we remove its electric charge and ignore its miniscule gravitational component. The part that is left is what we refer to as its *mechanical* component, which at this point is strictly a label. We know very little about this mechanical mass, except for the important fact that it constitutes 99.9% of the total mass of the electron. This is where the usefulness of an electron model becomes apparent. In quantum electrodynamics, the magnetic moment of the electron is calculated with incredible accuracy (Table 9.1). But QED has nothing definitive to say about W_E or W_H, since these quantities are swallowed up in the mass infinities that plague the theory. Thus the QED theorists, having no way of evaluating W_E or W_H, have no way to establish the existence or non-existence of a mass component W_M. Hence there is no discussion to be found in the literature about the properties of a non-electromagnetic mass. This is what led to Pais's lament, which was cited at the beginning of the chapter. However, the present electron model enabled us to draw specific conclusions about W_E and W_H, and these conclusions mandated the mass component W_M. Thus we are forced to consider what is in essence a new state of matter, whose properties *a priori* are completely unknown. *Mechanical mass* is a state of matter that seems to occur inside elementary particles, but which probably has no counterpart in the macroscopic world.

This mass problem is not limited to just the electron. From the point of view of spectroscopy, the *muon* appears as a "heavy" electron – an electron whose mass has been increased by a factor of 207, and whose radius has been decreased by this same factor. QED accounts for the magnetic moment of the muon (Table 9.2) in the same incredibly accurate manner as it does for the electron (Table 9.1), but it cannot account for the muon-to-electron mass ratio, nor even for the existence of the muon. The fact that the anomalous magnetic moment of the muon, like that of the electron, is about 0.1% of the total mass, indicates that the *mechanical* mass of the muon is increased by a factor of 207 over that of the electron, and the other quantities are scaled accordingly. The *tau* lepton, which appears as an even "heavier" electron, has a mass 17 times that of the muon, thus further compounding the mystery. Since the decays of the tau include both leptons and hadrons, this mass mystery spills over into the hadronic area, where it is manifested in the problem of the quark masses. The proton consists of two *up* quarks and one *down* quark. Spectroscopically, the *up* and *down* quarks appear quite similar to the electron, as we discussed in Chapter 5. Hence they presumably also have *mechanical* mass components. These mechanical mass components logically fit the *constituent-quark* picture (where the *up* and *down* quarks each have about 1/3 the proton mass), but not the *current-quark* picture (where the *up* and *down* quarks each have almost zero masses). The problem of the quark masses leads to the problem of calculating the proton-to-electron mass ratio, which is an unsolved problem that has bedeviled physicists ever since the discovery of the electron. The "mass problem" in physics may in fact be a "mechanical mass problem", and it cannot be properly addressed until we know what a mechanical mass is.

It attempting to determine the properties of the *mechanical* mass of a particle, we are starting with what is in essence a blank slate. How much can we actually find out? If our phenomenological guide – the electron model – is reliable, we can learn quite a lot. And the information we obtain is somewhat surprising.

The ***first*** conclusion we reach about this mechanical mass is that it is in the form of a *spatial continuum.* This conclusion is based on the properties of the relativistically spinning sphere described in Chapter 10. We started with a *uniform sphere of matter*, and then spun it relativistically. An integration over the mass components in the sphere yielded the relativistic moment of inertia $I = 1/2 \ mR^2$ (Eq. 10.6), and the consequent spin angular momentum $J = I\omega = 1/2 \ \hbar$. This integration contained the implicit assumption that the mass of the sphere is continuous. If it were "lumpy" on a scale comparable to the size of the electron, the calculated spin angular momentum would not agree with the observed value. Since we have no way at present of measuring any "lumpiness" or granularity inside the electron, the most straightforward assumption we can make is that the mechanical mass forms a continuum.

In the macroscopic world, the masses with which we familiar are clearly *not* continuous in the sense that we have just discussed. Consider a piece of iron, the stablest of all elements. The atomic nuclei inside this chunk of iron constitute 99.95% of the total weight, and yet they occupy only about 10^{-14} of the total volume. The atomic electrons, if taken to be Compton-sized particles, occupy about a millionth of the total volume. Thus iron consists mostly of empty space. All macroscopic masses are of this nature. There are no macroscopic mass distributions that fill the voids occupied by the vacuum state.

The ***second*** property of this mechanical mass that we can investigate is its *rigidity.* A spatially continuous mass may have a rigidity that is quite different from those of macroscopic masses. In iron, the forces that bind the atoms together are electromagnetic. If we examine the rigidity of a piece of solid iron, we are really examining the crystalline properties of an aggregate of electrostatically bound atoms. The deformation properties of iron arise mainly from the strength of the electrostatic forces, and the propagation speed for deformations is limited by the propagation speed of the electrostatic fields. Thus there is no reason to conclude that the rigidity of mechanical matter is similar to that of iron. But this brings up a interesting question: how do we measure the rigidity of a piece of mechanical matter? What we clearly must do is to apply a force and see what happens? However, there is a difficulty here. As we discussed in Chapter 7, the interactive forces in electron-electron scattering that arise from the charges on the electrons completely overshadow any interactive forces that arise from the mechanical masses. And we have no way of stripping the charge off an electron. But there two ways out of this impasse: *(1)* we can apply an *external* force to the *electric charge*, and then, by following the subsequent motion of the electron, deduce the manner in which this force has been transmitted from the charge to the mechanical mass; *(2)* instead of applying an external force, we can apply an *internal* force by rotating the electron and seeing how much it distorts. As we will see, both of these methods lead to the conclusion that the mechanical mass is exceedingly rigid. Let us first investigate the second possibility – rotations.

The rotational aspects of a particle mass represent another area (together with the nature of the mass itself) which has long been overlooked by physicists. Every spinning particle that we know about exists only in the spinning state. Thus it is not really necessary to inquire as to how the electron, for example, would appear if it were not spinning. An article in the journal *Physics Today* by Lev Okun (1989) on "The Concept of Mass" went into considerable detail on the manner in which a particle's "rest mass" (when it is not moving) should be distinguished from its "relativistic mass" (when it is moving). This article led to published responses from several readers.[2] However, none of these physicists considered the fact that the "rest mass" of a stationary spinning particle is in fact moving (spinning), and that it might be interesting to try and relate this non-moving "rest mass" to a non-moving and non-spinning "true rest mass". This turns out to be a useful topic, because it not only leads to the correct spectroscopic equations for the electron, but it also enables us to deduce something about the rigidity of the mechanical mass that forms the electron. The relativistic sphere equations that were derived in Chapter 10, which lead to the correct spin value for the electron, were obtained under the assumption that the spinning mass is continuous, as we discussed above. But they also contained the implicit assumption that the mass elements in the sphere are not distorted kinematically as a result of the rotational centrifugal forces. The envelopes of the integration mass elements were assumed to be invariant geometrically. Any distortion would have altered the value for the relativistic moment of inertia, and would have led to a calculated spin value in disagreement with experiment. But freedom from distortions means freedom from internal strains, and freedom from internal strains is a characteristic of a "rigid body" as defined in classical mechanics. In a rigid body, any internal perturbations are "instantaneously" distributed over the entire body, so that internal strains do not exist. Hence the fact that the spinning mechanical mass in the electron does not distort suggests that it functions as a rigid body – it is much more rigid than, for example, a piece of iron.

In considering the distortional effects of a relativistically spinning sphere, it should be kept in mind that such a sphere represents an unusual relativistic situation. As we discussed in Chapter 10, the *mass* of a sphere increases when it is spun relativistically. Its *envelope* is unchanged (the sphere is still a sphere), but its measured *volume* increases by the same factor as the mass increase (due to the non-Euclidean geometry inside the spinning sphere). Thus its measured *density* is unchanged. If density changes reflect internal strains, then the fact that the density is unchanged may be taken as an indication of the absence of internal strains.

Now consider the application of an external force to the electron. The rigidity of the electron determines the way it scatters. Turning this statement around, the way an electron scatters gives information as to its rigidity. An external electric force acts on the electron charge e, which, being in contact with the mechanical mass, produces a subsequent movement of the mechanical mass. In order to calculate this movement, we must know the manner in which the forces are transmitted through the mechanical mass. What we discover mathematically is that *if* the forces due to movement of the charge e are transmitted essentially instantaneously, then the *calculated* scattering (at most energies) is *point-like*. This is the result that occurs experimentally, and it indicates *aposteriori* that the mechanical mass of the electron is in fact *rigid*.

Let us assume that the mechanical electron mass fits the classical mechanics definition of a "rigid body". Thus perturbations are transmitted "instantaneously", and the internal strains are vanishingly small. This leads to a standard theorem in mechanics, Chasle's Theorem,[3] which we denote here as

THE GOLDEN RULE OF RIGID BODY SCATTERING

An external force applied to a rigid body can be separated into two components:

(1) a translational force that acts through the mass center;

(2) a torque that acts around the mass center.

Applying this *Golden Rule* to the present electron model, and making a classical calculation of the scattering of electrons or positrons off atomic nuclei (since they represent point-like scattering centers with respect to this large-electron model), we discover that at relativistic energies the scattering is indeed point-like. These calculations are described in detail in Chapter 16. Since this result is empirically correct, we use it to deduce the correctness of the Golden Rule, and hence the correctness of the assumption of rigidity for the mechanical mass of the electron.[4]

The ***third*** feature of this mechanical mass that we can deduce has to do with its interactions. As we discussed in some detail in Chapter 7, electron-electron scattering appears to be purely electromagnetic in nature. Electron-electron scattering was first calculated relativistically by Møller (1932), and the Møller scattering equation (Eq. 7.15) accurately describes both the *angular distribution* of the scattered electrons, and also the *absolute magnitude* of the cross section. The Møller equation is based on the Dirac theory of the electron, and it assumes that the interaction is purely electromagnetic. Thus if an electron consists, as we argue here, of a mechanical mass m with an embedded electrical charge e, it is the charge e that is giving rise to the entire measurable interaction between the electrons. The mass m, which supplies the inertial effects in the scattering, plays no significant role with respect to the interactive forces. In fact, since electrons at high energies are known to scatter off each another with very small impact parameters, high-energy electrons must literally pass through one another in the electron-electron scattering process. Electron scattering off atomic nuclei is considered in Chapter 16.

The ***fourth*** fact we can ascertain about the mechanical mass of the electron is that it in some manner carries the lepton quantum number. In the present electron model, all there is to the spin 1/2 electron is a mechanical mass that contains an embedded point charge e. Similarly, the spin 1/2 muon appears as a different (heavier) mechanical mass that contains the same charge e (see the muon-to-electron decay mode shown in Eq. 7.21). If we assume that the charges e in the electron and muon are in fact identical, then the electron lepton number and the muon lepton number (which characterize different lepton families) must be ascribed to the mechanical masses, and not to the charges.

With the relativistically spinning sphere model of the electron as a guide, we thus arrive at five conclusions about its *mechanical mass*:

(a) the mechanical mass constitutes 99.9% of the observed mass;

(b) the mechanical mass is spatially continuous;

(c) it functions internally as a "rigid body";

(d) it is essentially non-interactive with other particles,

(e) it carries the lepton quantum number.

Conclusion *(c)*, the rigidity of the mechanical mass, follows quite logically as a result of conclusion *(b)*, its spatial continuity. But conclusion *(d)*, its lack of external interactions, seems strange. It suggests that if we were to succeed in stripping the charge off of an electron, we would be left with a spin 1/2 particle that has no interactions with other particles. However, this is *not* a new concept in physics. This is precisely the property that is possessed by the spin 1/2 neutrino, which can traverse a hundred light years of lead before it interacts. And the similarity is even closer than that. The electron and the electron neutrino both carry the *electron lepton quantum number*. In the case of the electron, it seems clear that the lepton quantum number resides in the mechanical mass (see Eq. 7.22 and the accompanying discussion). In the case of the electron neutrino, the lepton quantum number must also be carried by the mechanical mass, since it is the only available carrier. But there is a complication in this analogy: the electron neutrino does not possess the same mechanical mass as the electron, and the muon neutrino does not possess the same mechanical mass as the muon. The calculated electron mechanical mass is 510,406 eV,[5] whereas an experimental upper limit of about 7 eV (Particle Data Group, 1992) has been established for the mass of the electron neutrino. Similarly, the calculated muon mechanical mass is 105,536 keV, whereas an upper limit of less than 270 keV (Particle Data Group, 1992) has been established for the mass of the muon neutrino. (It is evident from these mass values that the presence of the charge on an electron or a muon contributes in a fundamental way to the stability of its mechanical mass.) The really interesting experimental question with respect to the neutrino is the *lower* limit on its mass. Theoretically, the neutrino has been conventionally viewed as a massless particle, but recently grand unified theories (GUTS) have suggested that its mass might be finite. The present studies indicate that the spin of the neutrino arises from the rotation of a mechanical mass, in the same manner as does the spin of the electron.

This discussion of *mechanical masses* inside elementary particles may seem to be rather speculative, which indeed it is. But it must be kept in mind that the presence of such a mass component is clearly indicated theoretically, as Pais reminded us at the beginning of this chapter. The mass spectrum of the elementary particles continues to defy explanation. Gottfried and Weisskopf (1984, p. 100) have summarized this situation very forcefully:

"Unfortunately, QCD has nothing whatsoever to say about the quark mass spectrum, nor, for that matter, does any other existing theory."

And Richard Feynman (1985, p. 152), in a review of particle theory, voiced this same conclusion :

"Throughout this entire story there remains one especially unsatisfactory feature: the observed masses of the particles, *m*. There is no theory that adequately explains these numbers. We use these numbers in all our theories, but we do not understand them – what they are, or where they come from. I believe that from a fundamental point of view, this is a very interesting and serious problem."

It may be that the *mass* systematics of the elementary particles (Mac Gregor, 1990) is in essence a *mechanical mass* systematics. In the special case of the charged leptons, their mysterious mass intervals *must* come from their mechanical masses, since this seems to be the only feature that distinguishes an electron from a muon or a tau lepton. The neutrinos have their own mysterious mass intervals, which are compounded by the fact that their almost vanishingly small cross sections (10^{-42} cm^2) make it difficult to measure anything about them.

The conclusion from the present studies that the electron possesses a mechanical mass is not original. It has been considered at one time or another by a number of workers in the field. Rohrlich (1973, pp. 361-362) gives an interesting summary of this topic:[6]

> "... the classical theory of the electron ... has been dominated by the self-energy and stability problem and the associated question about the electromagnetic nature of the electron mass. Three types of solutions have been proposed.
>
> "The oldest solution proposed is that of Poincaré: the electron mass is partially electromagnetic and partially non-electromagnetic, the latter compensating the former .... .
>
> "Another very old idea ... is that of modifying either the Maxwell field equations or the equations of motion to attain a stable purely electromagnetic electron. ...
>
> "Finally, there is the simplest alternative, viz. to abandon any attempt at describing the mass of the electron within the framework of classical theory. In this approach the Coulomb self-interaction simply does not exist and there is no electromagnetic self-energy. To this class belong the theories by Fokker (1929) and by Wheeler and Feynman (1945, 1949) and my own theory (Rohrlich, 1964). The classical theory is here recognized as being entirely phenomenological with respect to both the mass and the charge of the electron. Clearly in such a theory there is no stability problem."

These ideas are described in some detail in Rohrlich's book *Classical Charged Particles* (1965). None of these studies include the concept of the relativistically spinning rigid sphere with its equatorial point charge. Although this sphere model appears to be unique with the present author, there is another model in the literature that also features a point electric charge and a Compton-sized non-electromagnetic mass. This is the "mixed model" of Mario Bunge (1955). Bunge derived Dirac-like equations for the motion of the electron which involve a *trembling* vector $\boldsymbol{x}$ and a *mean-position* vector $\boldsymbol{X}$. The vector $\boldsymbol{x}$ is identified with the oscillatory *zitterbewegung* motion of the electron, and also with the equations that characterize the Lorentz force on the electron. Hence Bunge identifies $\boldsymbol{x}$ with the position of a *point charge* on the electron. The vector $\boldsymbol{X}$ has a smooth non-oscillatory motion, and the average value of $d\boldsymbol{X}/dt$ is equal to the macroscopic velocity of the electron. Hence Bunge identifies $\boldsymbol{X}$ with the *center-of-mass* of the electron. He describes his mixed model as follows (Bunge, 1955):

> *"the charge is point-like while the mass is not"*;

and, in detail,

> *"the electron mass is spread over a region of dimensions of a Compton wavelength, a volume scanned by* $\boldsymbol{x}$ *and having no sharp boundaries".*

The Bunge model thus has certain similarities to the present sphere model, although lacking its detailed relativistic structure and its equatorial location for the charge.[7]

The material discussed thus far in the present book has consisted mainly of theoretical ideas and mathematical calculations. What we really need to make this viewpoint convincing is experimental evidence for the large size that we ascribe to the electron. This evidence may exist in Mott scattering, as we now go on to describe. In Chapter 16 we calculate the Mott scattering of electrons and positrons off atomic nuclei. Then in Chapter 17 we compare these calculations with experiment. At most energies the scattering is point-like. However, in the keV energy region there are experimental anomalies that may be telling us something.

Notes

1 The quotation shown on page 115 is from Kessler (1976, p. 206).

2 See the *Letters* section of *Physics Today*, May (1990), p. 13.

3 Chasle's Theorem is discussed in Goldstein (1950, pp. 124 and 143). Also see Synge and Griffith (1942, pp. 271-273 and 349-350).

4 For a brief comment on some of the problems associated with the assumption of rigidity, see Note 7 in Chapter 2. These problems arise in connection with the concepts of special relativity. If we extend our considerations into the domain of quantum mechanics, which we must eventually do in considering the properties of the electron, then the whole issue becomes more obscure. In particular, the problems associated with the Bell inequality seem to mandate some sort of nonlocality. Redhead concludes his book on *Incompleteness, Nonlocality, and Realism* with the following statement (Redhead, 1987, p. 169): "So there it is – some sort of action-at-a-distance or (conceptually distinct) nonseparability seems built into any reasonable attempt to understand the quantum view of reality. As Popper has remarked, our theories are 'nets designed by us to catch the world'. We had better face up to the fact that quantum mechanics has landed some pretty queer fish." If nonlocality is emerging as an essential feature of quantum mechanics, then it is transcending the bounds of conventional special relativity. Thus a rigid electron that also transcends those bounds may not be an entirely radical assumption.

5 The author cannot resist inserting a purely phenomenological result into this discussion. As mentioned in the text, the mechanical mass of the electron has the value $m_{mechanical} = m_{observed} \times (1 - \alpha/2\pi) = 510{,}406$ eV, with a negligible error. The mass difference between the negatively charged and neutral pions has been measured very accurately by Crawford *et al. (1988)* as 4,593,660 ± 480 eV. If we divide this mass difference by nine (see Figure 10.2, Eq. 10.12, and the accompanying text), we obtain 510,407 ± 53 eV (Mac Gregor, 1990, pp. 1040-1041), which accurately matches the mechanical mass of the electron.

6 The references in this quotation have been adapted to the present bibliography.

7 Papers delivered at the 1990 Antigonish Electron Workshop (Section D in Chapter 1) on the topic of zitterbewegung (see Note 8 in Chapter 1) contain discussions of electric charge motions that are similar to those of the Bunge (1955) model. Barut (1991, p. 109), in a continuation of the quotation that we cited on page 9 of Chapter 1, remarked as follows: "I want to describe a model in which a point charge performs as its natural motion a helix which gives an effective structure and size scale to the particle ... ."

Chapter 16

The keV Mott Helical-Channeling Window

Physicists today almost universally subscribe to the idea that the electron is truly point-like – a particle with no measurable dimensions. In the present studies we have gone to considerable lengths to set out the case for a different idea: that the electron is in fact much larger – Compton-sized – and that only its electric charge e is point-like. The virtue of the large electron model is that we can apply classical concepts to it, and thereby reproduce the spectroscopic properties of the electron. In order to establish the validity of this approach, we still have two important challenges to meet: one mandatory; the other highly desirable, but not mandatory. The mandatory challenge is the following:

can we demonstrate theoretically that this large electron scatters in a point-like manner, so that it is in agreement with experiment?

The other challenge is more broadly defined:

can we find any experimental indications for the large size that we here ascribe to the electron?

If the present large-electron model is in agreement with the scattering experiments that indicate point-like behavior, then it has some reasonable claim to existence, but it may not appear to be very useful. However, if it can also account for experimental results which are *not* consistent with point-like behavior, then it has a greatly enhanced plausibility. Thus far, no such experiments have been demonstrated to exist. We will try to meet both of these challenges.

What kind of experiment can we use to investigate possible finite-size effects stemming from this large electron? In the case of electrons that are *bound* in atomic orbitals, the quantum-mechanical tipping of the spin axis relative to the axis of quantization serves to cancel out finite-size effects, as we discussed in Chapters 7 and 13. Specifically, the electric quadrupole moment vanishes at the quantum-mechanically prescribed spin angles $\pm\theta_{QM}$ (Eq. 7.8), since the rotating electric charge forms a *quantum current loop* (Eq. 7.10). This suggests that we should investigate the interactions of *unbound* (free) electrons, where this type of cancellation may not occur.

The simplest *free-electron* experiment that we can devise is elastic scattering from a localized scattering center. The smallest practical scattering center (other than another electron, which from the present viewpoint is not so localized) is an atomic nucleus. The smallest and lightest atomic nucleus is the proton, which has a measured electromagnetic radius of less than 10^{-13} cm (see Table 5.1), and hence is smaller by almost three orders of magnitude than the present Compton-sized electron, which has a radius $R \sim 7 \times 10^{-11}$ cm. The heavier atomic nuclei have measured electromagnetic radii of only a few fermis, and thus are still very small as compared to the size of the large electron. The *classical* elastic scattering of a charged particle (positive or negative) from an atomic nucleus is denoted as *Rutherford scattering*, and the angular distribution of the scattering is given by the "relativistic Rutherford equation"[1]

$$\sigma_R(\theta) \;=\; \left(\frac{Ze^2}{m_O c^2}\right)^2 \frac{(1-\beta^2)}{4\beta^4 \sin^4(\theta/2)} \;, \tag{16.1}$$

where θ is the scattering angle, Z is the charge on the nucleus, m_O is the rest mass of the electron, and $\beta = v/c$ is the relativistic velocity of the electron. The Rutherford equation can be derived from the Schrödinger equation (Mott and Massey, 1949, Ch. III), or directly from the $1/r^2$ Coulomb force law (French, 1958, App. V). *Quantum-mechanical* elastic nuclear scattering is denoted as *Mott scattering*. Mott scattering differs from Rutherford scattering in that it is derived from the relativistic Dirac equation (Mott and Massey, 1949, Ch. IV) rather than the non-relativistic Schrödinger equation, and it includes electron spin effects that are not contained in the Rutherford formulation. The Mott scattering cross sections are quite similar to the Rutherford cross sections of Eq. (16.1), and they are conventionally expressed as deviations from the corresponding Rutherford cross sections (Sherman, 1956). In addition to the spin effects that distinguish Mott cross sections from Rutherford cross sections, the original Mott formalism has subsequently been extended to include screening effects of the atomic electrons, finite nuclear size effects, and (in the case of positrons) annihilation processes. All of these effects are in general quite small, and the deviation of Mott scattering cross sections from Rutherford scattering cross sections is usually less than a factor of two. The Rutherford cross sections themselves vary by more than four orders of magnitude in going from 10° to 180° in scattering angle.

The electric charge *e* and magnetic moment μ of the electron are responsible for its electrostatic and magnetic interactions with atomic nuclei. In general, electrostatic forces are much stronger than magnetic forces. In the present discussion, which is centered on Mott scattering, we will consider only the dominant electrostatic effects. From the standpoint of electrostatic interactions, the difference between a large electron and a point-like electron is that the equatorial charge on a large spinning electron traces out a *helical* path as the electron moves along, whereas the charge on a point-like electron coincides with the center-of-mass trajectory. These trajectories are displayed in Figure 16.1. If the helical motion that is shown in Figure 16.1 affects the electron trajectory, then we have a manifestation of the finite size of the large electron; i.e., if initially identical large-electron and point-electron mass-center trajectories lead to different scattering angles, then the finite size of the electron becomes an observable phenomenon.

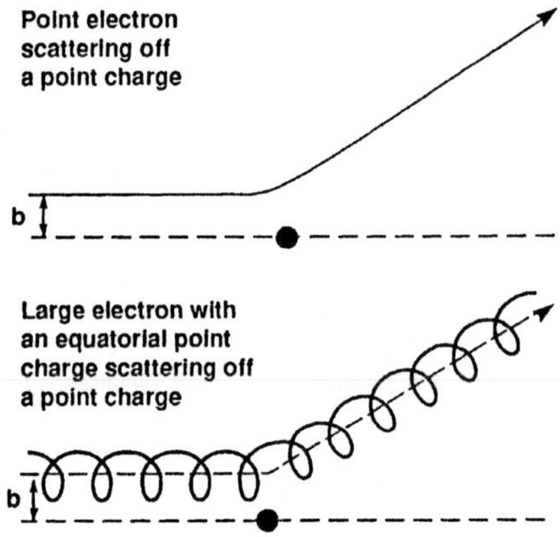

Fig. 16.1. A comparison of point-electron and large-electron trajectories, showing the motion of the electric charge.[2] The displacement distance b is the asymptotic impact parameter.

In Mott nuclear scattering, if the radius of the helical motion is much smaller than the "effective scattering radius", the effect of this helical motion will be very small. The effective scattering radius is determined at low energies by the range of the Coulomb forces that mediate the scattering. As we described above, atomic nuclei have much smaller dimensions than a Compton-sized electron, and hence smaller dimensions than the helical path shown in Figure 16.1. However, the Coulomb forces at low energies are so strong relative to the momentum of the incoming particle that an incoming electron or positron does not "see" the true nuclear size. It is deflected long before it gets anywhere near the nucleus. Thus the effective size of the scattering center at low energies is very large – much larger than the actual size of the nucleus. If the particle energy is so low that this effective size is large as compared to the radius of the helical path, then the helical motion will have essentially no effect on the scattering. This effect serves to impose a *low energy limit* on the observability of finite-size effects in Mott scattering.

The "effective size" of a scattering center can be specified in terms of the range of asymptotic impact parameters that are important in the scattering. The *impact parameter* is the parallel distance that an incoming particle trajectory is displaced from a trajectory aimed at the center of the nucleus. It is the distance b shown in Figure 16.1. If the interaction were turned off, it would correspond to the "distance of closest approach" to the nucleus. Let us specify, for example, a measurement of the differential elastic scattering cross section over an angular range from 10° to 180°. The impact parameters that correspond to 10° scattering and 180° scattering then define the pertinent impact parameter range. Electrons and positrons, which experience attractive and repulsive nuclear Coulomb forces, respectively, have different actual distances of closest approach to the nucleus for the same initial impact parameter. However, for low energies and large impact parameters, where a Born approximation type of scattering is valid, the relationship between impact parameters and scattering angles is essentially the same for both of these particles.

In the case of Rutherford scattering, we can calculate the asymptotic impact parameters analytically (French, 1958, App. V). But we can make these results more realistic by carrying out a numerical integration of *point-particle* trajectories on a computer. If we do the simplest type of numerical calculation, using the $1/r^2$ Coulomb force and a fixed particle mass, we obtain the same cross sections and corresponding impact parameters as are given by Eq. (16.1). However, by doing a numerical integration, we can add in screening effects and nuclear size effects, and the mass of the particle can be relativistically adjusted at each step of the iteration. (An electron may have a low initial velocity, so that relativistic effects are unimportant, but it gets accelerated to relativistic velocities as it approaches the nucleus.) Several thousand iteration steps are required for an accurate trajectory, and several thousand trajectories are used to map out a range of impact parameters for each incident energy. Thus accurate numerical integrations are time-consuming. The calculations for *helical* trajectories are even more time-consuming. Each step in a helical calculation requires transformations among several sets of direction cosines, and a large volume of space must be systematically sampled over a range of relatively small impact parameters. (The calculations described in the present chapter required several weeks of full time operation on a battery of workstation-type computers.[3])

In Figure 16.2 we show the asymptotic impact parameters b that were calculated for *positron* point-particle scattering on carbon, copper, and gold nuclei. (Positron impact parameters are similar to electron impact parameters, and wide-angle positron trajectories are easier to calculate accurately.) The calculations were for pure Coulomb scattering,

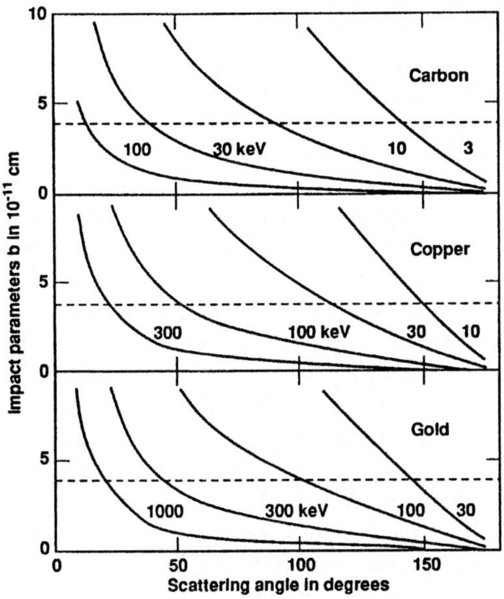

Fig. 16.2. Asymptotic impact parameters b for point-like positron scattering off carbon, copper and gold nuclei. The dashed lines denote the positron Compton radius, R_C.

with screening effects, nuclear size effects,[4] and relativistic mass effects included. The abscissa in Figure 16.2 shows the scattering angle, and the curves give the impact parameters b in units of 10^{-11} cm for various incident energies. If the curves fall below the dashed lines that correspond to the Compton radius, then we have $b < R_C$, so that the helical motion for large-positron or large-electron trajectories is large in comparison to the impact parameters, and hence in comparison to the "interaction size". But if the curves lie above the dashed lines, then the helical motion is small in comparison to the interaction size. As can be seen in Figure 16.2, the curves for carbon at 3 keV, copper at 10 keV, and gold at 30 keV lie above the dashed lines for all but the widest scattering angles. Thus these energies represent *low energy thresholds* for the onset of finite-size effects in Mott scattering. At lower energies, the Coulomb "effective interaction range" is much larger than R_C, and hence much larger than the radius of the helical path followed by the charge on a large positron or electron.

We have now established approximate lower limits for the energies at which finite-size effects might be observable in the Mott scattering of positrons and electrons off various atomic nuclei. But if we go to energies that are above these thresholds, *what do we expect to find?* How is the effect of *helical* charge trajectories manifested in Mott scattering? In order to give a definitive theoretical answer to this question, we would have to carry out quantum-mechanical Mott scattering calculations that include these helical trajectories. This has not been done, and it is probably not necessary at this stage of the investigation. The question we are asking is really a classical one. The effect of the helical motion on Mott scattering cross sections should be roughly the same as the effect of the helical motion on Rutherford scattering cross sections, and we can see intuitively what this effect is. Let us select scattering events in which the asymptotic impact parameters for wide-angle scattering, b_{wide}, are comparable to the electron (or positron) Compton radius R_C (see Figure 16.2). We then distinguish two cases: *(a)* $b_{wide} > R_C$; *(b)* $b_{wide} < R_C$. These two cases are illustrated in Figure 16.3.

For the *point-like* Rutherford scattering shown in Figure 16.1, each scattering event occurs in a two-dimensional plane that is defined by the impact parameter b and the initial trajectory. The scattering plane is fixed. For the *large-electron* or *large-positron* scattering shown in Figures 16.1 and 16.3, however, the helical path followed by the charge on the particle represents a *continuous rotation of the scattering plane.* The effect of this rotation is to partially cancel out the transverse components of the Coulomb force. The strongest transverse cancellation is produced at small impact parameters (case b in Figure 16.3). The Coulomb force in this case is roughly constant over the full 360° rotation of the scattering plane. Thus a transverse force component in (say) the $+y$ direction is compensated 180° later in the rotational cycle by a similar force component in the $-y$ direction. This cancellation of the transverse forces leads to a *reduction* in the wide-angle scattering cross sections, and hence an *increase* in the small-angle scattering. The large electron or large positron is *channeled* into forward scattering angles. This is the process that we refer to as *Mott channeling* or *helical channeling*. For larger incident impact parameters (case a in Figure 16.3), the transverse cancellation is much smaller, and the main effect is an increase in the average $1/r^2$ Coulomb force. For *repulsive* positron Coulomb scattering

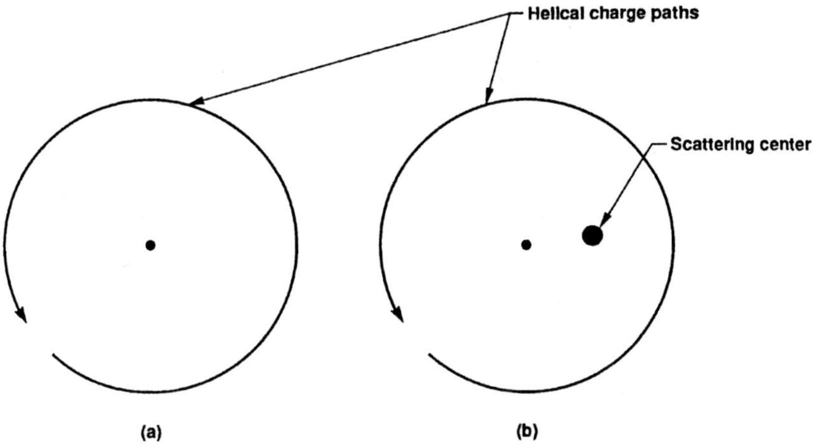

Fig. 16.3. Asymptotic impact parameters for the helical charge trajectories (Figure 16.1) of electrons incident on a localized Coulomb scattering center (an atomic nucleus). In *(a)* the scattering center lies outside of the incident helix, and in *(b)* the center lies inside the helix. In each case the helical motion represents a continuous rotation of the scattering plane. In case *(b)* the rotation of the scattering plane encompasses a full 360°, and there is a substantial cancellation of the transverse Coulomb forces over a cycle of helical motion; this cancellation leads to enhanced forward scattering – to *channeling*. In case *(a)* the transverse cancellation is much smaller, and the main effect is an enhancement of the average $1/r^2$ Coulomb force. For positrons, this enhanced force leads to increased wide-angle scattering – to *anti-channeling*; for electrons, it leads to increased channeling.

off an atomic nucleus, the increased force in case *(a)* leads to an *increase* in the wide-angle scattering, so that we actually have *anti-channeling*. For *attractive* electron Coulomb scattering, on the other hand, the increased force that occurs in case *(a)* aims the helical trajectory closer to the scattering center, thus tending to change a case *(a)* trajectory into a case *(b)* trajectory. Hence we expect to find larger channeling effects for electrons than for positrons, and also somewhat lower thresholds.

Since Rutherford cross sections (Eq. 16.1), and hence also Mott cross sections, fall off steeply (by four orders of magnitude) at wide angles, an effect that produces a large decrease in the wide-angle scattering will lead to a very small increase in the corresponding small-angle scattering. Hence a *helical channeling* process that causes (say) a 10% reduction in wide-angle scattering will have no measurable effect on forward scattering. We now have an answer to the question raised above as to the effect in Mott scattering of the helical motion of the charge on the electron:

The helical charge motion that occurs in large-electron Mott scattering will, above a certain energy threshold, lead to a reduction in the wide-angle elastic scattering, and it will have essentially no effect on the small-angle elastic scattering.

We have seen that *below* a certain energy threshold, which is approximately a few keV, the helical charge path associated with the trajectory of a large electron has no effect on Mott scattering cross sections. *Above* this threshold, the helical charge motion causes a reduction in the wide-angle elastic scattering. What happens if we move to even higher energies? When we do this, we discover an effect that imposes an *upper energy limit* on this Mott channeling process:

> *In order for helical channeling to occur, the incident electron or positron must make several revolutions of the spin axis during the "interaction time" of the scattering event.*

The channeling effect arises because transverse Coulomb forces during one portion of the spin cycle are partially canceled out 180° later in the cycle. But if the scattering event happens so swiftly that there is no time for several spin revolutions to occur, then this cancellation effect cannot take place, and there will be no channeling. As we go up in energy, several factors conspire to decrease the number of spin revolutions that occur during the interaction time: *(1)* the impact parameters *b* decrease with increasing energy (Figure 16.2), so that the entire scattering takes place at very short distances, thus shortening the interaction time; *(2)* at relativistic energies, the time dilatation slows the particle spin rotational frequency as observed in the laboratory frame of reference; *(3)* the particle is fore-shortened relativistically, which also serves to decrease the interaction time. Thus at high energies there is no channeling effect, and the scattering of a large electron essentially goes over into the scattering of a large *non-rotating* electron. It is not clear *a priori* what the scattering of a large non-rotating electron looks like, but we will demonstrate by direct calculations that, with the assumptions used here, it is point-like in nature. In particular, the assumption of *rigidity* for the mechanical mass of the electron is the crucial element that leads to this result.

We have now achieved a qualitative picture of the Mott channeling process for finite-sized electrons or positrons. At low energies (below 1 keV) there is no channeling effect, and the scattering is point-like. After the energy threshold for a particular nucleus has been reached (in the multi-keV region), there is a decrease in the observed wide-angle scattering. At relativistic energies (in the MeV region) the channeling disappears, and the scattering again becomes point-like. We can make these remarks quantitative, or at least semi-quantitative, by carrying out computer calculations of large-electron and large-positron Rutherford scattering. The particle model we use is the one that was displayed in Figure 14.1; namely, a large non-interacting sphere of mechanical matter that has an embedded equatorial point charge *e*. The mechanical matter supplies all of the inertial effects for the electron, and the electric charge *e* supplies the Coulomb interactions with the atomic nucleus. The physical coupling between the charge *e* and the mechanical mass is specified by *The Golden Rule of Rigid Body Scattering* defined in Chapter 15. If we assume that the mechanical mass fulfills the classical definition of a "rigid body", then it does not have internal strains, and any applied external force (from the charge *e*) can be decomposed into a *translational force* that acts through the center-of-mass plus a *torque* that acts as a couple around the center-of-mass. As the particle moves along, the task of following the motion of the electric charge requires a transformation from a set of direction cosines fixed in the revolving particle to a set of spatially-fixed direction cosines at each iteration step. Also, the energy of the particle has to be continuously monitored, so as to make the proper relativistic corrections for time dilatation, fore-shortening, and mass variations. And the task of computing torque effects requires cross products involving the various force components. Thus the computing time for each particle trajectory is quite large.

The computational procedure that was followed was to calculate a set of *point-particle* trajectories using a range of initial impact parameters, and then calculate a matching set of *large-particle* trajectories using the same range of impact parameters. The calculated cross sections for these two types of scattering can then be directly compared, and any deviations from the point-like values serve as indications of finite-size effects. A difficulty with the large-particle calculations is that the scattering depends not only on the initial impact parameters, but also on the initial spin and charge configurations. In order to properly sample the interaction space, a whole set of spin-charge combinations has to be used for each value of the impact parameter. In a standard calculation, 180 different initial spin axis orientations were uniformly distributed over a sphere, and 30 initial charge positions around the equator were used for each spin orientation, so that 5,400 large-particle trajectories were calculated for each value of the impact parameter. The large-particle calculations for a single atomic nucleus and a single bombarding energy typically involved several hundred thousand trajectories.

The computer calculations that we have just outlined are for classical Rutherford-type scattering. These calculations are adequate to reveal the general behavior of *helical* scattering as contrasted to *point-particle* scattering, but they do not yield reliable cross section *spectra*, since the various refinements that enter into Mott scattering have not been taken into account, and the effects these refinements produce are comparable to the expected channeling effects. Thus the information obtained from the present calculations is best expressed in terms of integrals over wide angular ranges. A useful way to characterize the magnitude of the channeling effect is to calculate the ratio $R_{F/P}(\theta_0)$ of the integrated wide-angle cross sections for *finite-size* and *point-particle* scattering, where

$$R_{F/P}(\theta_0) = \int_{\theta_0}^{180°} \sigma_{finite}(\theta)d\Omega \div \int_{\theta_0}^{180°} \sigma_{point}(\theta)d\Omega . \tag{16.2}$$

The computed values of $R_{F/P}(\theta_0)$ turned out to be quite similar for values of the lower limit θ_0 ranging from 10° to 30°. Thus the value θ = 20° was adopted as a standard. The calculated $R_{F/P}(20°)$ ratios for electron and positron scattering off various nuclei are shown in Figure 16.4. On the basis of the impact parameters displayed in Figure 16.2, we had estimated that the channeling thresholds for C, Cu and Au would be roughly 3, 10 and 30 keV, respectively. As Figure 16.4 demonstrates, these thresholds are approximately correct. The *magnitudes* of the channeling effects displayed in Figure 16.4 are not reliable, since they depend on the detailed spectral shapes, which are not given accurately in this classical Rutherford-type calculation. However, the *locations* of the thresholds and the *sign* of the effect should be the same for both Rutherford and Mott scattering.

Electrons are attracted to the positively charged atomic nucleus, and thus have smaller "effective impact parameters" than do positrons, which are repelled by the nucleus. Hence, as we discussed in connection with Figure 16.3, we expect the threshold for electron channeling to occur at a lower energy than it does for positron channeling, and we expect

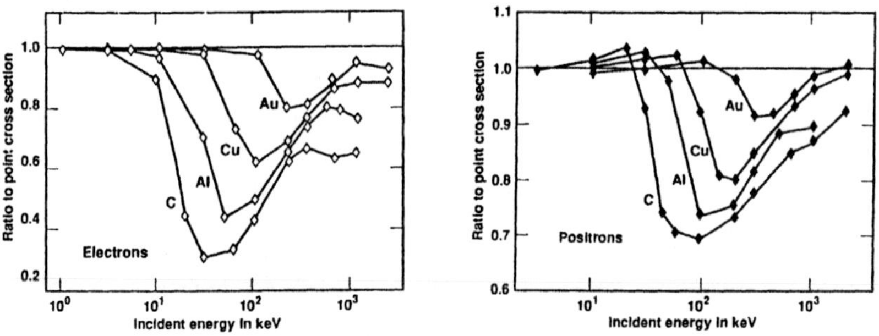

Fig. 16.4. Calculated $R_{F/P}(20°)$ ratios (Eq. 16.2) for electron and positron scattering off carbon, aluminum, copper and gold.

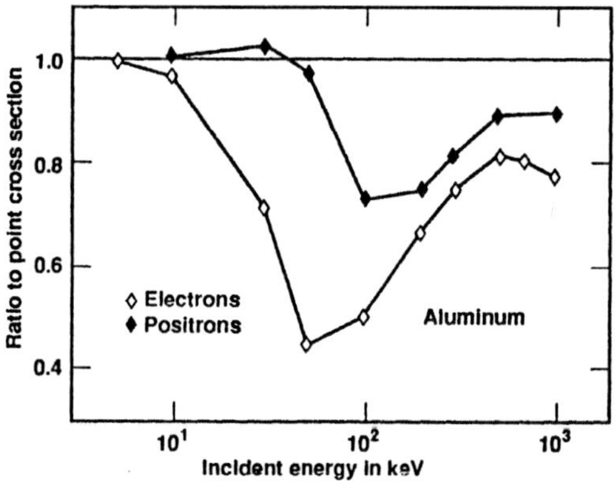

Fig. 16.5. A comparison of the electron and positron $R_{F/P}(20°)$ ratios for aluminum. Electron helical channeling occurs at a lower threshold.

the channeling effect to be somewhat larger for electrons than for positrons at the same energy. In order to illustrate this, we display in Figure 16.5 the electron and positron $R_{F/P}(20°)$ ratios for aluminum taken from Figure 16.4.[5] Figures 16.4 and 16.5 illustrate the effects of the case *(a)* and case *(b)* helical trajectories defined in Figure 16.3. The *positron* wide-angle scattering in these figures first increases and then decreases as the incident energy is raised above 1 keV. This is caused by a shift from case*(a)* to case *(b)* impact parameters as the energy is increased. The *electron* wide-angle scattering, on the other hand, shows only a decrease, since the effect of the enhanced attractive Coulomb force in case *(a)* is to steer the electron more strongly into case *(b)* scattering channels.

The fall-off of the channeling effect at high energies shown in Figures 16.4 and 16.5 should be considered to be only qualitatively correct, since it depends in part on relativistic factors which were only approximately taken into account. Also, calculations demonstrated that the torque effect arising from the action of the external electrostatic field on the equatorial charge *e* has a rather small effect on the $R_{F/P}$ ratios (see Table 16.1 below), and it was generally not included in the calculations, since it greatly increased the running time of the problems.

As mentioned above, the large-electron and large-positron scattering states for the $R_{F/P}$ ratios calculated here were based on an *isotropic* distribution of spin axis orientations. When these calculations were repeated with the electrons and positrons in definite *helicity states* (where the quantum-mechanical projection of the spin axis is either parallel or antiparallel to the center-of-mass motion), approximately the same results were obtained.

In the high-energy limit of large-electron or large-positron scattering, the scattering process is essentially that of a *non-rotating* particle, since the rotational motion during the very brief interaction time is inconsequential. However, *this is not the same as point-particle scattering, because the center-of-mass of the large particle does not coincide with the location of the charge e.* According to the Golden Rule of Rigid Body Scattering, the action of this force is through the center-of-mass, where it is transmitted "instantaneously" from the charge *e*. This is the type of scattering that imposes the greatest demands on our assumption of the mechanical mass as a "rigid body". As we have emphasized in Chapter 15, the *mechanical* matter that constitutes the bulk of the electron is a type of material that does not exist macroscopically. Hence it is difficult to imagine exactly what properties it might have. We know from special relativity that electromagnetic forces in a vacuum propagate at the velocity of light, *c*. And we know that the forces inside of (for example) a lump of iron are electromagnetic. Thus it is clear that internal stresses in iron cannot be transmitted faster than *c*. Gravitational forces also have this same limitation. The propagation speed of stresses in mechanical matter, which we assume here to be infinite, is a topic that is beyond the scope of the present studies. Calculationally, we have approximated the high-energy scattering limit for large electrons or large positrons by turning off the rotational motion and ignoring torque effects. Table 16.1 shows representative $R_{F/P}(20°)$ ratios that are obtained for *(a)* a complete large-positron scattering calculation, *(b)* the same calculation with torque effects removed, and *(c)* the same calculation with torque effects removed and the rotational motion set equal to zero. As can be seen in Table 16.1,

TABLE 16.1. $R_{F/P}(20°)$ RATIOS FOR POSITRON SCATTERING CALCULATIONS

Element	*Energy (keV)*	*Complete calculation*	*Torque excluded*	*Torque excluded and zero rotation*
carbon	100	0.74	0.68	1.01
copper	300	0.86	0.84	1.02
gold	300	0.95	0.91	1.00

the torque effects, while not negligible, are relatively small, and are comparable to other uncertainties in the calculations. The rotational effects, however, are very decisive. When they and the torque effects are removed, the calculated scattering is accurately point-like. The bombarding energies that we actually employed for the results displayed in Table 16.1 are (for calculational reasons) quite low, so the procedure we have just described for estimating high-energy behavior is not rigorous. The calculational techniques used in the present work are not well-suited for high energy problems (above 1 MeV), which require a careful relativistic treatment. Thus we have used these modifications to low-energy calculations in order to mock up the expected behavior at high energies. Hence the results displayed in Table 16.1 should be considered as merely suggestive.

There is another phenomenon that can also contribute to anomalous electron and positron scattering in the keV range. This is the *zitterbewegung* effect that is known to occur for bound-state electrons, and which we discussed in Chapter 7. As an electron travels toward an atomic nucleus, it can interact either with the vacuum state or else with the Coulomb field of the nucleus so as to emit and then reabsorb virtual photons. The recoil effects from this process cause a "trembling" motion of the electric charge. In the case of bound electron orbitals, the average position of the charge is spread out by the zitterbewegung effect over a region of space that is comparable to the Compton radius R_C of the electron. In the case of unbound electron orbitals, the magnitude of the zitterbewegung spreading can only be determined by a rigorous QED calculation, which was not carried out. The zitterbewegung effect on Mott scattering can be crudely modeled in computer calculations by introducing *ad hoc* perturbations, in a manner similar to the treatment of the helical trajectories discussed above, but it turns out in actual calculations to depend critically on the *magnitude* of the perturbations at very short distances.[6] In terms of Figure 16.3, it depends on whether the zitterbewegung perturbations are large enough to produce case *(b)* channeling effects, or are so small that they produce only case *(a)* enhancements or no effect at all. The magnitude of the zitterbewegung effect may depend markedly on the separation distance between the electron or positron and the atomic nucleus, and also on the collision dynamics. We can model the helical trajectories of the large electron quite accurately, because they represent an essentially classical situation. But the zitterbewegung perturbations are not *a priori* well-determined in a quantitative sense, and the quantitative features are crucial to the calculation. Thus detailed theoretical studies are required before the zitterbewegung contributions to Mott scattering can be estimated with any accuracy. In the case of the large-electron model, we will in principle have both helical spirals and superimposed zitterbewegung perturbations. And, in principle, we will also have a spreading out of the electrical charge due to vacuum polarization effects, which may serve to obscure some of the helical motion.

The present model of a large electron was formulated on the basis of its ability to mathematically reproduce the *spectroscopic* features of the electron, including in particular its spin and magnetic moment. This is a task that remains beyond the purview of point-electron theories. But does this large electron reproduce the *dynamical* features of the electron – in particular, its scattering cross sections? When we analyze the Mott scattering properties of this large electron, as we have done here, we discover that the large electron mimics a point electron at energies below 1 keV, and probably also above 1 MeV. However, in the crucial region between 1 keV and 1 MeV, it leads, for both electrons and positrons, to a predicted *reduction* in the observed wide-angle elastic Mott scattering cross sections: the helical motion of the charge on the electron or positron produces a *helical channeling* effect that enhances the forward scattering. The question which then arises is a crucial one: *are there experiments that show signs of this helical channeling effect?* This is the topic we take up in the next and final chapter. The studies that we have done here are theoretical, and the fate of all theories rests ultimately in the hands of the experimentalists. We will see that there are some experimental results which are suggestive of helical channeling, but the main fact which emerges is that the experimental region in question has not been adequately explored.

Notes

1. The "relativistic" Rutherford equation shown in Eq. (16.1) differs from the original Rutherford equation by the factor of $(1 - \beta^2)$ in the numerator, which changes the rest mass m_0^2 in the denominator into the relativistically moving mass.
2. For helical-looking electron trajectories that are based on the *zitterbewegung* motion of the point charge e in a large Compton-sized mechanical electron, see Bunge (1955, Fig. 1).
3. The author is grateful to Dr. Roger White of Lawrence Livermore National Laboratory for making SUN SPARC STATION computational facilities available for these calculations.
4. Although finite nuclear size corrections were applied, they are negligible at the low energies used here.
5. The results shown in Figures 16.2, 16.4 and 16.5 have been published in Mac Gregor (1992).
6. Rather extensive zitterbewegung computer calculations were carried out by the author, but they turned out to depend so strongly on the unknown zitterbewegung magnitudes at very short distances that the results were judged to be inclusive, and are not presented here.

Chapter 17

Experimental Evidence for Helical Channeling

Mott scattering, the elastic scattering of electrons or positrons off atomic nuclei, is a mature subject in physics, and it has been reasonably well investigated both experimentally and theoretically. Thus it seems there should be no surprises lurking in these experiments. From a theoretical point-of-view, there has been no reason to go searching in particular energy regions for discrepancies between experiment and Mott theory. However, this does not necessarily mean such discrepancies don't exist. As Kessler (1976, p. 206) noted in the quotation shown at the beginning of Part IV, fields of physics which had been considered to be thoroughly investigated and well understood often contain unexpected results, sometimes with extensive ramifications. The present author did an experimental thesis on nuclear beta decay, which was carried out in the years just prior to the discovery of parity non-conservation (Mac Gregor, 1952, 1954). At that time, no one was even considering beta decay events which did not conserve parity, but they were quickly discovered once the motivation to look for them had been supplied. This motivation, it should be noted, came from the $\theta - \tau$ puzzle of kaon decays in particle physics, and not from nuclear physics. The moral here is that, when searching for experimental discrepancies, it is helpful to have theoretical guidance.

Using the calculations of Chapter 16 as our guide, we now search among the various Mott scattering experiments in an attempt to locate anomalies that are suggestive of finite-size effects in the electron. These anomalies can occur in Mott single scattering experiments, where they take the form of *helical channeling*. Or they can occur in Mott double scattering experiments, where they take the form of *helical depolarization*. As we will see, there is some evidence to be found in both of these areas.

A. The Search for Helical Channeling

The task of reproducing the spectroscopic properties of the electron led us to the relativistically spinning sphere model of the electron, which consists of a Compton-sized mechanical mass and an embedded equatorial point charge. When we calculated the manner in which this large electron scatters from an atomic nucleus (Chapter 16), we discovered that at most energies the scattering appears to be point-like. However, at energies where the Compton radius of the electron is comparable to the asymptotic impact parameters that characterize the scattering, there seems to be an observable effect: *the wide-angle elastic scattering cross sections are too low.* This effect arises from the helical motion of the charge on the electron (Figures 16.1 and 16.3), which partially cancels the transverse components of the scattering force and thus *channels* incoming particles into forward scattering angles. An inspection of positron *point-particle* impact parameters for Rutherford scattering (which is qualitatively similar to Mott scattering) indicates that this helical channeling effect should have a *low-energy limit* of a few keV for low-Z atomic nuclei, and a few tens of keV for high-Z nuclei (Figure 16.2). A *high-energy limit* of perhaps one MeV is imposed on this helical channeling process by the fact that at the higher energies the scattering interactions are very localized and the "interaction times" are very short, so that the spiraling motion (which is slowed relativistically) becomes insignificant.

Calculations of the Rutherford scattering of Compton-sized electrons and positrons confirm this expected helical channeling behavior (Figures 16.4 and 16.5). These calculations also indicate that the effects are larger for low-Z nuclei than for high-Z nuclei (Figure 16.4), and are larger for electrons than for positrons (Figure 16.5). Thus if the present ideas have validity, we expect to find Mott elastic scattering anomalies appearing in an energy region that extends from roughly 10 keV up to a few hundred keV, and to occur most strongly in low-Z nuclei. The expected form of the helical channeling anomaly, as mentioned above, is a *decrease* in the observed *wide-angle* elastic Mott scattering relative to the calculated Mott scattering. Now, the Rutherford and Mott elastic scattering cross sections are peaked very sharply in the forward direction (Eq. 16.1), so that the wide-angle cross sections represent only a very small fraction of the total elastic cross section. Thus the small-angle cross sections (which correspond to much larger impact parameters) are essentially unaffected by perturbations in the wide-angle scattering. Hence *small-angle* scattering is *not* expected to exhibit helical channeling effects. It should be noted that the *magnitude* of the helical channeling effect is not reliably given by the calculations described in Chapter 16, since these are for classical *Rutherford* scattering, and not for the actual quantum mechanical *Mott* scattering. Also, the shape of the high-energy cut-off that occurs at relativistic electron or positron energies is probably not reliably given. Thus these Rutherford calculations serve mainly as a guide in looking for actual experimental discrepancies in the Mott scattering data.

Historically, the experimentalists who study Mott-type scattering have fallen into two quite distinct groups. The *atomic* physicists use electron and positron scattering primarily to investigate the properties of the electron cloud which surrounds the atomic nucleus, and their very extensive experiments have been centered mostly below 1 keV. The *nuclear* physicists focus their attention on the interactions of electrons and positrons with the atomic nucleus itself, and their experiments, which initially extended down into the keV region, have long ago moved up to higher energies. Comprehensive angular measurements of thin-solid-target Mott differential cross sections in the multi-keV region have more-or-less ceased a quarter of a century ago. Thus the *wide-angle* Mott experiments in the multi-keV region that serve as tests of channeling effects are quite fragmentary, and many of them were not done with the present refinements in instrumentation. More recently, experimentalists have moved from solid to gas targets in order to eliminate multiple-scattering effects. However, gas targets introduce the problem of absolute target calibration, which usually requires comparison methods. With these gas targets, the atomic physicists have extended their measurements up well into the keV region, but their efforts have been directed mainly at *small-angle* scattering, where the scattering is strongly influenced by the electron cloud that surrounds the nucleus. The one notable exception is mercury, which is easily produced as a gaseous target, and which has been well-studied at multi-keV energies over the whole angular range. Electron scanning microscopes use beam energies in the multi-keV region, but the scattering is at very small angles.

A computer-assisted literature search was carried out to find the published papers on electron and positron Mott elastic scattering, and a compilation was made that includes the multi-keV experiments[1] together with a representative sampling of the voluminous atomic-domain experiments.[2] This compilation is graphically illustrated in Figure 17.1, and it reveals a clear separation in energy between these two types of experiment.

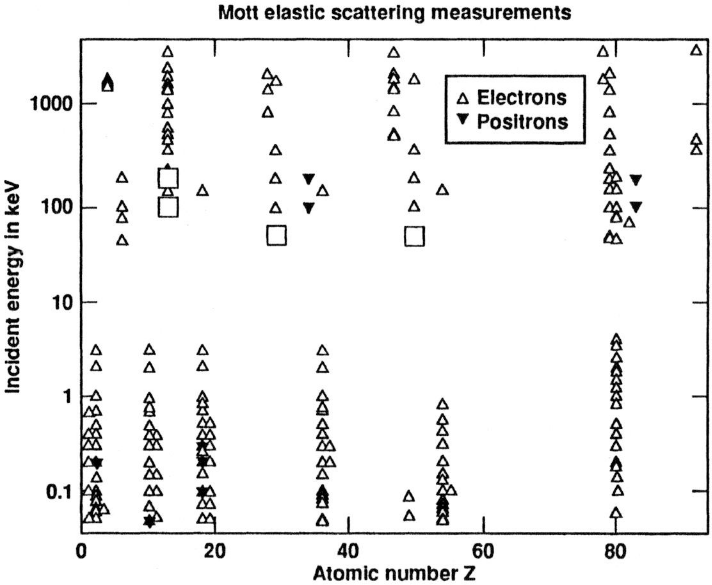

Fig. 17.1. An energy plot of Mott wide-angle thin-target elastic scattering measurements. All of the multi-keV experiments carried out after 1950 are included,[1] together with a representative sampling of the low-energy atomic-type experiments.[2] The four open squares denote electron data sets in which channeling effects may be in evidence.

As a guide to the reader, we have listed in Appendix A a compilation of papers that relate to Mott *single-scattering* experiments. These deal mainly with differential cross sections for elastic scattering, and they form the basis for the data plot of Figure 17.1. Appendix B contains a similar compilation of papers that relate to Mott *double-scattering* experiments. These deal mainly with polarization and asymmetry measurements. Figure 17.1 is an energy plot of selected data sets from Appendix A, with the energy of the incident particle plotted against the atomic number of the scattering nucleus. It is restricted to experiments carried out after 1950, since the earlier measurements were not very accurate. It is also restricted to experiments that serve as tests of Mott helical channeling effects – namely, differential cross section measurements that feature the use of thin targets, and that have angular ranges which extend to wide angles. As can be seen in Figure 17.1, there is an energy gap extending from about 5 to 45 keV where no modern experiments exist that are suitable for testing Mott channeling effects. This is precisely the energy region where we expect the helical channeling thresholds to appear. The Mott measurements below 5 keV and the majority of the multi-keV measurements show no evidence of helical channeling. However, there are four data sets which do exhibit anomalies – two in aluminum (Z = 13) at 100 and 200 keV, one in copper (Z = 29) at 50 keV, and one in tin (Z = 50) at 50 keV. These are the electron measurements labeled in Figure 17.1 with large squares. This anomalous behavior occurs for relatively low-Z nuclei, and at the lowest energies used in each experiment. Perhaps significantly, these anomalies appear in the most recent and most accurate thin-target experiments carried out on these particular nuclei.

We now describe these experimental anomalies in more detail. The experimental situation with respect to Mott scattering in the multi-keV region has been well-summarized by the experimenters themselves. In a comprehensive review article, which also included their own experimental results, Spiegel *et al.* (1959) discussed the Mott elastic differential cross section measurements done up to that time and commented that

"none of the previous experiments meet all the criteria necessary to test the validity of the Mott theory".

Their own experiments (Spiegel, 1959), which were for electron scattering off aluminum, nickel, silver and gold nuclei at energies of 1, 1.75 and 2.5 MeV, gave results that were generally below the Mott values by a few percent.

A subsequent experiment on aluminum at 1 MeV was carried out by Rester and Rainwater (1965a), and was then expanded (1965b) to include ten bombarding energies extending from 0.1 to 3 MeV, and a range of foil thicknesses extending down to 11 $\mu g/cm^2$. Their experimental results show evidence of a channeling type of anomaly. This is illustrated in Table 17.1, which gives average values for each of the ten Rester and Rainwater differential cross section sets, expressed as ratios of the experimental values to the exact Mott unscreened theoretical values of Doggett and Spencer (1956). The assumption was made (Rester and Rainwater, 1965b) that screening corrections are unimportant. This assumption was later justified by the screening calculations of Zeitler and Olsen (1966), which extend down to 200 keV, and which demonstrate that the screening effects for aluminum are in fact negligible. As Table 17.1 shows, the experimental Rester and Rainwater aluminum cross sections are in good agreement with theory for all energies

TABLE 17.1. EXPERIMENTAL MOTT ELASTIC SCATTERING CROSS SECTIONS FOR ALUMINUM (RESTER AND RAINWATER, 1965b), EXPRESSED AS RATIOS TO THE EXACT THEORETICAL MOTT UNSCREENED CROSS SECTIONS OF DOGGETT AND SPENCER (1956).

Energy (keV)	*Average deviation from Mott*	*Number of points*
100	-10.2%	8
200	-8.6%	7
500	+0.1%	6
700	-4.2%	8
1000	-2.1%	9
1250	+2.4%	6
1500	+1.2%	6
2000	+6.1%	9
2500	+6.9%	6
3000	+5.8%	6

from 500 to 1,500 keV. However, at 2, 2.5, and 3 MeV, the cross sections are *higher* than the theoretical values by an average of about 6%, and hence are in contradiction with the aluminum results of Spiegel *et al.* (1959) cited above. But the really interesting results for our present purposes are at the lowest energies, 100 and 200 keV. Here the cross sections are significantly *lower* than the theoretical values, by averages of 10.2% and 8.6%, respectively, and hence might be exhibiting helical channeling effects. The present literature search indicates that this discrepancy has not been resolved. The Rester and Rainwater experiments, which were published in 1965, appear to be the last thin-target multi-keV electron scattering experiments carried out on aluminum. The experiments performed later were focused on plural and multiple scattering effects in relatively thick foils,[3] and on backscattering effects from metal surfaces.[4]

Although there has been no direct follow-up on the Rester and Rainwater aluminum measurements, an indirect check was made by Kobayashi and Shimizu (1972). These workers commented that the Rester and Rainwater measurements constitute

"one of the most reliable experiments for the scattering of 0.1 - 3.0 MeV electrons by Al".

They noted that

"the measured values of the cross section at 0.1 and 0.2 MeV are apparently lower than the Mott cross section calculated by Doggett and Spencer (1956) for pure Coulomb field. The disagreement cannot be explained by calculations taking into account the screening effect of atomic electrons (Lin, 1964). There may be other unknown causes responsible for this disagreement. This effect may possibly depend on the charge of incident particles. Therefore, it is interesting to attempt measurements with positrons."

Then, remarking that

"no reliable measurement has so far been performed for elastic scattering of positrons of energies lower than 300 keV", they proceeded to carry out a positron scattering experiment, using energies of 100 and 200 keV that match the lowest Rester and Rainwater energies, but using selenium and bismuth targets rather than aluminum. The results they obtained were in general agreement[5] with their own theoretical calculations (which included screening corrections). Thus they found no evidence for a scattering anomaly that depends on the *sign of the charge* of the bombarding particle.

Aluminum is the most-studied low-Z element from the standpoint of keV Mott scattering experiments. The other two elements that are indicated in Figure 17.1 as exhibiting channeling-type experimental anomalies are copper and tin. Accurate measurements on copper, tin, and gold at 50, 100, 200, and 400 keV were carried out by Motz, Placious, and Dick (1963). In a later paper devoted specifically to keV Mott scattering experiments that serve as tests of wide-angle screening corrections, Kessler and Weichert (1968) characterized the Motz, Placious, and Dick measurements as

"up to now the most accurate experiment".

In the Motz (1963) paper, their experimental results for copper and gold were compared to the matching Mott screening calculations of Lin (1964). The results for tin were not similarly handled, since Lin did not calculate screening corrections for tin. However, later screening calculations by Zeitler and Olsen (1966) at 200 and 400 keV included copper, tin, and gold. These calculations demonstrated that, to a fairly good approximation, the screening corrections for tin are halfway between the screening corrections for copper and gold. This assumption can be used, together with the Lin (1964) screening values for copper and gold, to obtain interpolated screening corrections for tin at energies below 200 keV. The average Motz, Placious, and Dick experimental differential cross section values for copper, tin, and gold, expressed as ratios to the corresponding screened Mott cross sections, are shown in Table 17.2. The values for tin that are displayed in Table 17.2 are

TABLE 17.2. MOTT SCATTERING CROSS SECTIONS FOR COPPER, TIN, AND GOLD (MOTZ, PLACIOUS, AND DICK, 1963), EXPRESSED AS RATIOS TO THE MOTT SCREENED CROSS SECTIONS OF LIN (1964) FOR COPPER AND GOLD, AND LIN (1964) (INTERPOLATED) FOR TIN.

Average deviation from the exact Mott screened cross sections

Energy (keV)	*Copper*	*Tin*[6]	*Gold*
50	-13.2%	-9.6%	-0.3%
100	+0.4%	+0.8%	+4.1%
200	+0.2%	-2.4%	+1.0%
400	-4.9%	+1.0%	-3.8%

based on the interpolation procedure described above.[6] Motz *et al.* (1963), commenting only on copper and gold, remarked that the agreement is good

"except for Cu at 50 keV where the discrepancies are not understood".

Lin (1964), who estimated that his screened cross sections were accurate to 1%, echoed this assessment:

"The only exception is the case of copper at the electron energy of 50 keV. The reason for such a big disagreement is not clear."

As Table 17.2 shows, the differential cross section ratios for copper at 50 keV are low by an average of 13%, whereas the ratios at the higher energies are close to unity. This well-recognized anomaly in copper at 50 keV, like the anomalies in aluminum at 100 and 200 keV described above, has not been resolved. The only subsequent thin-target multi-keV measurement on copper was made at 98.6 keV (Grachev, 1978), and it yielded relative elastic differential cross sections which follow the general shape of the Mott cross section.

We can see from Table 17.2 that the 50 keV channeling-type anomaly which Motz *et al.* (1963) and Lin (1964) reported for copper also occurs for tin. The differential cross section ratios for tin at 50 keV are low by an average of about 10%, whereas the ratios at the higher energies agree with theory. Since the tin screening corrections at 50 keV were not calculated directly, but were obtained by an interpolation between the copper and gold screening calculations at 50 keV, as described above, it might be argued that the observed discrepancy for tin at 50 keV is attributable to a faulty interpolation procedure. However, if this were the case, then we would expect most of the discrepancy between experiment and theory to show up at the forward scattering angles, where the screening corrections are the largest. But it is the *back* angles in tin at 50 keV that exhibit the largest discrepancies: the experimental cross section values at 30 and 40 degrees are each about 8% low, whereas the five cross section values between 50 and 90 degrees are all between 10% and 11% low. These observed discrepancies are larger than any errors that can logically be attributed to the screening interpolation procedure. Furthermore, this same interpolation procedure carried out at 100 keV yields a very accurate fit to experiment. These measurements on tin have never been repeated, so the tin anomaly at 50 keV, together with the copper anomaly at 50 keV and the aluminum anomalies at 100 and 200 keV, stand as the latest and probably most accurate measurements carried out on these elements in this energy range. These are the four large squares shown in Figure 17.1.

Although the experimental copper (Z = 29) and tin (Z = 50) Mott differential cross sections at 50 keV exhibit the anomalous channeling-type behavior that is displayed in Table 17.2, no such effect is seen in the gold (Z = 79) cross sections. These are in accurate agreement with theory at all energies. This indicates that the channeling effect, if it actually exists, is most prominently a low-Z phenomenon. This conclusion was suggested by the computer calculations displayed in Figure 16.3, and it receives confirmation from precision measurements of Mott electron scattering on mercury that were carried out by Kessler and Weichert (1968) at energies of 46, 79, 100 and 204 keV. The mercury target was in the form of an atomic beam of mercury atoms. This technique has the advantage that the beam is effectively "thin" to electrons, so that multiple scattering effects in the target are eliminated. It has the drawback that the density of target atoms cannot be accurately measured directly, but must be deduced from the scattering off a known comparison beam. The ratio of the two scatterings then gives the mercury cross sections. The comparison beam used for the Kessler and Weichert measurements was hexadecane ($C_{16}H_{34}$). Theoretically calculated cross sections were used for carbon and hydrogen, and the assumption was made that the atoms in the hexadecane molecule constitute independent scattering centers, so that their scatterings can be linearly combined. The experimental differential cross sections that were obtained for mercury agree within the experimental error of about 2% with the Mott scattering calculations of Lin (1963, 1964), Bühring (1968), and Moore and Fink (1972). These cross sections show no signs of the low-energy anomalies exhibited by the aluminum, copper and tin nuclei. The Kessler and Weichert (1968) measurements for mercury are thus in agreement with the Motz, Placious, and Dick (1963) measurements for gold in demonstrating that channeling-type anomalies are not in evidence in high-Z nuclei at energies of 50 keV and above. However, the Kessler and Weichert results have a further ramification. They suggest that the theoretical cross section values used for hydrogen and carbon in the hexadecane calibration beam are correct, so that these very-low-Z materials do not exhibit helical channeling effects at energies of 50 keV and above. This conclusion seems incompatible with the aluminum results of Rester and Rainwater (1965b) and the copper and tin results of Motz, Placious and Dick (1963), and it brings up the question of data consistency, which we now discuss.

The multi-keV Mott elastic differential cross section measurements that are displayed in Figure 17.1, and which are documented below,[1] include all of the wide-angle thin-target experiments published in 1950 or later that the author was able to find. These experiments reveal the four anomalies we have described in detail above. However, as is demonstrated in Tables 17.1 and 17.2, the measurements that were made on these same elements at slightly higher energies by the same experimental groups (Rester and Rainwater, 1965b; Motz, Placious and Dick, 1963) do not exhibit these anomalies. Also, the anomalies do not show up for high-Z elements, and there is the indirect evidence of Kessler and Weichert (1968) that they are not showing up in the lowest-Z elements. Thus, independently of any interpretation that is attributed to these anomalies, *the multi-keV measurements that are displayed in Figure 17.1 do not seem to constitute a self-consistent data* set. In addition to the anomalous behavior that occurs in the 50 to 100 keV range, the Rester and Rainwater (1965b) aluminum measurements at 2 MeV and above are inconsistent with the measurements of Spiegel *et al.* (1959), which indicates that there may also be problems at higher energies. In order to accurately delineate channeling anomalies that appear as 10% effects in wide-angle elastic scattering, we require further thin-target experiments on low-Z nuclei. In particular, the energy region below 50 keV, where we might expect to see rather dramatic helical-channeling threshold effects, needs to be explored.

As Figure 17.1 shows, no modern thin-target wide-angle Mott elastic scattering measurements have been carried out in the energy region between 5 and 45 keV. However, a variety of other measurements have been completed in this energy domain. Physical chemists and atomic physicists use multi-keV small-angle scattering on gas targets to probe the structure of the electron cloud that surrounds the nucleus.[7] This scattering can be unfolded to reveal electron correlations in the atomic orbitals.[8] Electron microscopists use very-small-angle electron scattering in the multi-keV range,[9] and large-angle (back-angle) multi-keV scattering is utilized for sample thickness determinations and for Auger electron microscopy.[10] Plural and multiple scattering effects have been extensively studied throughout this energy region.[11] Thus the paucity of Mott thin-target wide-angle elastic scattering experiments in the 5 to 45 keV region seems to stem more from a lack of theoretical motivation than from any intrinsic difficulties in carrying out these measurements (although multiple scattering difficulties do become more severe at the lower energies). Hopefully, the present studies will help to provide this motivation.

B. The Search for Helical Depolarization

Our goal in this chapter has been to discover experiments that exhibit signs of helical channeling, and hence furnish evidence for finite-size effects in the electron. The only type of experiment we have discussed thus far is elastic scattering from atomic nuclei. However, there is another feature of Mott nuclear scattering that is relevant, and which may be equally useful in the search for finite-size effects. This is the polarization asymmetry that occurs when an electron or positron scatters from the nucleus. This asymmetry is detectable in a second scattering. The polarization is produced by the interaction of the magnetic moment of the electron with the magnetic field that is created by the motion of the atomic nucleus as viewed in the rest frame of the electron.[12] The Mott scattering that occurs for a polarized electron beam is given by the equation

$$\sigma(\theta,\phi) \quad = \quad I(\theta)\,[1+S(\theta)\,\mathbf{P}\cdot\hat{\mathbf{n}}] \quad .$$

In this equation, **P** is the polarization of the incoming electron along the direction $\hat{\mathbf{n}}$, and $S(\theta)$ is the *Sherman function* (Sherman, 1956)[13] that characterizes the interaction with the nucleus. If the initial electron beam is unpolarized, then the scattering cross section is $\sigma(\theta) = I(\theta)$, which is the case we have been considering up to now. But if the initial beam is polarized, then the scattering cross section has an azimuthal dependence on ϕ that gives an observable effect in a second scattering. This is where finite-size effects come into play. If the electron is large, then the helical motion of its charge (Figure 16.1) represents a continual rotation of the scattering plane that defines the azimuthal axis. This rotation tends to wash out part of the azimuthal dependence, and hence tends to *decrease* the observed Mott polarization. We have *helical depolarization.* Unfortunately, we cannot use a classical approximation to theoretically investigate Mott double scattering, as we used classical Rutherford scattering to approximate Mott single scattering, because this classical scattering formalism does not include spin effects. Thus we have no direct theoretical guidance in searching for helical depolarization effects. But the Mott energy window that delineates the energy range where the helical motion of the charge is of importance must be roughly the same for Mott *polarization* experiments as it is for Mott *elastic scattering* experiments.

In the multi-keV energy region the Sherman function S is larger for high-Z nuclei than it is for low-Z nuclei, since the high-Z nuclei produce stronger magnetic fields and hence stronger electron magnetic moment interactions. Thus Mott high energy asymmetry measurements have been performed mainly on high-Z nuclei. But at low energies, where the Mott interactions are more complicated, large polarizations can also be found in low-Z nuclei. Experimentally, the asymmetry measurements that have been carried out divide into two energy groups. The *multi-keV* group extends from 10 keV on up,[14] and it features solid gold targets almost exclusively, since these are conventionally used in Mott polarimeters. The *low-energy* group extends from a few keV down to the low eV range,[15] and it features mainly gas targets so as to eliminate multiple-scattering effects. These gas targets include the noble gases as well as vaporized high-Z elements, so the low-energy measurements encompass the range of Z-values from low to high.

Guided by the systematics of Mott elastic scattering cross sections, we expect helical depolarization anomalies to show up at energies above 10 keV. Since essentially all of the Mott asymmetry measurements above 10 keV are for gold targets,[14] we must look at the Sherman S-values for gold to see if multi-keV anomalies exist. These anomalies consist of experimental S-values that are low in comparison to the corresponding screened Mott S-values as calculated by Lin (1964) or by Holzwarth and Meister (1964). The search for helical depolarization effects is complicated by the fact that multiple scattering becomes important in the low multi-keV region, right where we expect to find these helical effects, and multiple scattering also lowers the observed Mott asymmetries. As we will see, the asymmetry measurements on gold do exhibit a persistent anomaly below 100 keV which seems to be larger than can be attributed to multiple scattering, and which may be indicative of helical depolarization. We now describe these experiments.[14] Since the Sherman function $S(\theta)$ reaches a maximum absolute value in gold at an angle of about 120° for energies near 100 keV, the value $S(120°)$ is often quoted for comparison purposes.

The early Mott asymmetry measurements of Ryu (1952, 1953), Pettus (1958), and Nelson and Pidd (1959) contained no multiple-scattering corrections, and the experimental S-values they obtained for gold were considerably below the corresponding theoretical values, especially at the lower energies. The subsequent experimental measurements of Bienlein *et al.* (1959) and Apalin *et al.* (1962) took multiple scattering effects into consideration, but these measurements also yielded S-values at the lowest energies (120 keV and 45 keV, respectively) that were below the calculated S-values (Lin, 1964). Lin commented that these low values

> "cannot be attributed to screening effects alone. The source of discrepancy may still lie in incomplete treatment of the plural and multiple scattering effects."

Later experiments of Mikaelyan, Borovoi and Denisov (1963) and Van Klinken (1966) gave S-values for gold which were in agreement with one another, and which were low at energies below 100 keV as compared to the calculated S-values of Lin (1964) and Holzwarth and Meister (1964) (see Fig.10 of Van Klinken, 1966). Van Klinken made the following comment:

> "The disagreement of the present data with theory is most obvious at low energies (below 100 keV). Even though the experimental errors are largest in this region, we assume that there exist fundamental theoretical shortcomings for the calculated curves."

A measurement by Eckardt, Ladage and Moellendorff (1964) gave an S-value at 100 keV that was in general agreement with theory (Fig. 1 of Eckhardt). A thick target experiment by Boersch, Schliepe and Schriefl (1971) utilized an energy filtering technique to remove multiple-scattering effects, and it gave results that were in general agreement with Mikaelyan (1963), and hence were low at energies below 100 keV (see Fig. 7 of Boersch).

More than a decade later, two modern Mott asymmetry experiments on gold continue to show this low-energy anomaly in the gold S-values. Gray, Hart, Dunning and Walters (1984) made measurements over an energy range from 20 to 120 keV. They obtained S-values that agree with the theory of Holzwarth and Meister (1964) at energies above 60 keV, but which are too low at energies below 60 keV (see Fig. 5 of Gray, 1984). The following evaluation was made by these authors:

"At electron energies ≥ 60 keV the present data are in good agreement with the theoretical values. However, at lower energies the data fall somewhat below the calculated values. This discrepancy may result from incomplete rejection of inelastically scattered electrons or from multiple elastic scattering in the foil. However, as evident from Fig. 5, the present zero energy-loss values of S_{eff} are significantly larger than those obtained by Mikaelyan et al. in a double scattering experiment."

They thus suggest that their multiple scattering corrections may in fact be adequate, and that the observed anomaly below 60 keV may represent a significant departure from theory.

An experiment by Campbell, Herman, Lampei and Owen (1985) was carried out at about the same time as the Gray (1984) experiment. It gave relative S-values for gold which, when normalized to Holzwarth and Meister (1964) at 100 keV, follow the shape of the Holzwarth curve down to low energies (see Fig. 9 of Campbell). However, Campbell *et al.* used an energy-loss extrapolation technique that was later questioned by Fletcher, Gay and Lubell (1968, p. 921).

The final experiment we cite here is that of McClelland, Scheinfein and Pierce (1989). These workers, who made measurements for both thorium and gold over an energy range from 10 to 100 keV, obtained results for gold in agreement with those of Gray (1984) (see Fig. 3 of McClelland). They commented as follows:

"Also shown in Fig. 3 for comparison are results from Gray *et al.* The agreement with the present results for gold is excellent. Furthermore, the disagreement with the theory of Holzwarth and Meister at lower energies found by Gray *et al.* is confirmed by our measurements."

McClelland *et al.* did not refer to the work of Campbell (1985).

We can see from these experiments that there is a persistent anomaly in the asymmetry measurements for gold. The experimental values for the Sherman function S are in reasonable agreement with the screened calculations of Lin (1964) and Holzwarth and Meister (1964) for electron energies above roughly 100 keV, but they have been consistently too low at energies well below 100 keV. The range of energies measured extends down to 10 keV, and thus spans the energy region where we might expect helical depolarization effects to appear. It has conventionally been assumed (without proof) that the low S-values for gold below 100 keV can be attributed solely to multiple scattering effects. However, it would be much more interesting if what we are actually seeing here is a manifestation of finite-size effects in the electron. But a precautionary comment is in order. If we are in fact observing helical depolarization, then we might expect to find evidence for a low-energy threshold below which the anomalously low S-values are no longer observed. Now, measurements on mercury gas targets at energies of 2 keV and below give S-values which are in good agreement with theory,[16] so the helical depolarization threshold for high-Z elements must lie *above* 2 keV. But the solid-gold-target measurements of Gray (1984) and McClelland (1989), which extend down to 20 and 10 keV, respectively, show no signs of coming back into agreement with theory at these low energies. Thus uncorrected multiple scattering effects are certainly important at 10 to 20 keV, and the question arises as to whether they constitute all of the observed anomaly, or just part of it. The theoretical calculations themselves (Lin, 1964; Holzwarth and Meister, 1964) need to be modified in order to take multiple scattering effects into account. If it can be established that at (say) 50 keV, the persistent anomaly observed in the gold Mott asymmetry measurements is larger than can be accounted-for by multiple scattering phenomena, then there will be a real incentive to pursue these ideas in more detail.

C. Final Remarks

We have now come to the end of *The Enigmatic Electron.* If an author is writing a fictitious mystery story, he is duty-bound to bring all of the parts together at the end of the story and resolve everything to the satisfaction of the reader. However, if he is chronicling a real-life mystery, he may not be able to accomplish this. The electron itself is a scientific mystery story of the first rank. Is it a point, or is it very large? Is it an object that can be classically understood, or is it purely quantum mechanical? Is it a particle, or a wave, or both, or alternately one or the other? Do the anomalies that appear in Mott scattering in the multi-keV region represent the first tantalizing signs of a finite size for the electron, or are they merely manifestations of trivial multiple scattering effects?

The protagonist in *The Enigmatic Electron* is easy to identify. It is the *relativistic spinning sphere.* Without this concept, there would be no viable spectroscopic model for the electron, no way of classically visualizing its properties, and no book. There would be no purpose in using anything more than a totally abstract representation for the electron in terms of its field operators. Thus the present book is really written to extol the virtues of the relativistic spinning sphere, which represents a natural extension of the concepts of relativity – both special *and* general, since they both lead to Eq. (10.1). The fact that this model has not found its way into the standard body of physical knowledge, at least as a heuristic device, seems to the present author to be rather puzzling. It would seem that physicists have an obligation to examine a problem from all points of view, especially if it remains an unsolved problem. The electron does admit a classical – or at least semi-classical – representation, and we all ought to be aware of this fact.

Suppose that the Mott single scattering and double scattering experiments in the multi-keV region are carefully investigated, and they turn out to exhibit no signs of *helical channeling* or *helical depolarization.* Does this invalidate the concept of the large electron? The answer is clearly *no!* We know that free electrons in a Stern-Gerlach experiment do not reveal their half-integral spins: the Lorentz forces that arise from the charge on the electron overwhelm the magnetic splittings of the two spin orientations.[17] Similarly, there may be effects (e.g., vacuum polarization) which operate in Mott scattering to cancel out contributions due to the helical motion of the charge. But if the electron is truly large, and if it always operates in such a way as to cancel out its finite size, then we may legitimately inquire if there is any point in paying attention to its size. If the electron is always contained inside an inscrutable black box, we may as well work with just the properties of the box. Hopefully, the Creator has not been that unkind to us, and has given us an electron we can eventually understand.

Notes

1 The following is a listing of *multi-keV* (> 10 keV) wide-angle experiments that serve as tests of Mott elastic scattering theory: Neher (1931); Klarmann and Bothe (1936); Cox and Chase (1937); Barber and Champion (1938); Champion and Barber (1939); Chase and Cox (1940); Randels, Chao and Crane (1945); Van de Graaff *et al.* (1946); Buechner *et al.* (1947); Lyman, Hanson and Scott (1951); Kinzinger and Bothe (1952); Paul and Reich (1952); Anthony, Waldman and Miller (1953); Kinzinger (1953); Bayard and Yntema (1955); Chapman *et al.* (1955); Ruane, Waldman and Miller (1955); Spiegel, Waldman and Miller (1955); Brown, Matsukawa and Stewardson (1956); Damodaran and Curr (1956); Pettus, Blosser and Hereford (1956); Kessler (1959); Spiegel *et al.* (1959); Keck (1962); Motz, Placious, and Dick (1963); Rester and Rainwater (1965a,b); Kessler and Weichert (1968); Kobayashi and Shimizu (1972); Grachev *et al.* (1978, 1980).

2 The following is a representative listing of medium-energy (< 10 keV) *electron* experiments that serve as tests of Mott elastic differential cross section scattering theory: Kessler and Lindner (1965); Bromberg (1969, 1974); Gupta and Rees (1975); Williams and Willis (1975a); Williams (1975b); Williams and Crowe (1975c); DuBois and Rudd (1976); Jansen *et al.* (1976a); Jansen and de Heer (1976b); Saha, Chaudhuri and Ghosh (1976); Williams, Trajmar and D. Bozinis (1976); Bransden and McDowell (1978) (summary paper); Buckman, Noble and Teubner (1979); Gupta and Mathur (1979); Klewer, Beerlage and van der Wiel (1980); Vuskovic, Maleki and Trajmar (1984); Hyder *et al.* (1986); Register, Vuskovic and Trajmar (1986); Holtkamp, Jost, Peitzmann and Kessler (1987); Iga *et al.* (1987); Mitroy, McCarthy and Stelbovics (1987); Rao and Bharathi (1987); Danjo (1988); Peitzmann and Kessler (1990); Marinkovic *et al.* (1991).

The following is a listing of *medium-energy* (< 10 keV) *positron* experiments that serve as tests of Mott elastic differential cross section scattering theory: Coleman and McNutt (1979); Hyder *et al.* (1986); Kauppila *et al.* (n.d.); Kauppila and Stein (1987); Kauppila and Stein (1989).

3 See Oms, Erman and Hultberg (1969); Erman (1970).

4 See Bronshtein and Pronin (1976a,b); Gergely (1987).

5 The Kobayashi and Shimizu experimental results were, on the average, 1.2% low and 0.7% high for Se at 100 and 200 keV, and 4.0% low and 1.7% low for Bi at these same energies, so that *"the experimental cross sections tend to be a little lower than the theoretical ones"* (Kobayashi and Shimizu, 1972). However, the quoted experimental uncertainties were about ± 6% for each data point, and hence were quite large in comparison to the discrepancies shown in these averages.

6 If the Zeitler and Olsen screening corrections for tin at 200 and 400 keV (Ref. 7) are used in Table 17.2 instead of the extrapolation procedure described in the text, the average values of these ratios become -0.1% and +2.0%, respectively, instead of -2.4% and +1.0%.

7 See Fink and Kessler (1966); Fink and Bonham (1969); Wellenstein, Bonham and Ulsh (1973); Fink and Moore (1977); Duguet, Bennani and Roualt (1983); Ketkar, Fink and Bonham (1983); McClelland and Fink (1985).

[8] See Peixoto, Bunge and Bonham (1969); Naon and Cornille (1972); Naon, Cornille and Kim (1975).

[9] See, for example, Burge and Smith (1962).

[10] Kanter (1964); Cosslett and Thomas (1965); Niedrig and Sieber (1971); Schmoranzer, Grabe and Schiewe (1975); Bronshtein and Pronin (1976a,b,c); Ichimura, Aratama and Shimizu (1980); Gergely (1987); Lesiak, Jablonski and Gergely (1990).

[11] Nigam, Sudaresan and Wu (1959); Scott (1963); Cosslett and Thomas (1964a,b); Braicovich, De Michelis and Fasana (1967); Oms, Erman and Hultberg (1969); Erman (1970).

[12] An excellent presentation of the Mott polarization formalism is given in Kessler (1976).

[13] As an historical sidenote, the present author and Professor Sherman were teammates for three years on the University of Michigan graduate physics department softball team.

[14] The following is a listing of *multi-keV* (> 10 keV) asymmetry measurements, mostly on gold targets: Shull, Chase and Myers (1943); Ryu (1952a,b), (1953); Pettus (1958); Nelson and Pidd (1959); Beinlein *et al.* (1959a,b); Greenberg *et al.* (1960); Spivak, Mikaelyan, Kutikov and Apalin (1961); Apalin *et al.* (1962); Mikaelyan, Borovoi and Denisov (1963); Eckardt, Ladage and Moellendorff (1964); Van Klinken (1966); Braicovich and De Michelis (1968); Van Duiden and Aalders (1968); Boersch, Schliepe and Schriefl (1971); Hodge *et al.* (1979); Gray *et al.* (1984); Campbell *et al.* (1985); Fletcher, Gray and Lubell (1968); Tang *et al.* (1988); McClelland, Scheinfeld and Pierce (1989).

[15] The following is a listing of *low-energy* (<10 keV) asymmetry measurements, mostly on gas targets: Deichsel and Reichert (1964), (1965); Jost and Kessler (1966); Eckstein (1967); Loth (1967) (solid targets); Mehr (1967); Schackert (1968); Bartschat, Hanne, Wolcke and Kessler (1981); Berger and Kessler (1986); Kelley (1989) (review article); Geesmann, Bartsch, Hanne and Kessler (1991).

[16] The low-energy Mott asymmetry measurements have been reviewed by Kessler (1969).

[17] See Kessler (1976, Sec. 1.2).

Postscript

Pandorean Lessons from Special Relativity

Physics is the study of physical phenomena – the study of the "real world" – at the simplest and most elementary level. Mathematics is the language that is used to quantitatively describe these phenomena. The way physicists tend to picture the development of physics is that the *experimentalists* first work out the essence of what is going on, and then the *mathematical physicists* come along and re-express these findings in terms of elegant and often abstract equations. First comes the *physics*, and then the *mathematics*. And, indeed, many if not most discoveries have proceeded in this manner. A classic example is provided by the field of electromagnetism. Michael Faraday and others empirically deduced the strange workings of the electric and magnetic fields, and then James Clerk Maxwell summarized their results in the set of elegant equations that bears his name.

Sometimes, however, the process works in the other direction: the mathematical equations are deduced before the physicists have figured out what is really happening. An early example was Kepler's discovery that planets move in elliptical – not circular – orbits. Kepler had the mathematics, but not the physics. It remained for Isaac Newton to relate the elliptical motion to the $1/r^2$ gravitational force law. More recently, the Ritz combination principle and the Rydberg constant together yielded an undeniably accurate reproduction of atomic spectra, but it took the Bohr atom to give an explanation for these equations. Another example was provided by Alfred Landé, who wrote down an interval rule that accounted for the anomalous Zeeman effect in the alkali atoms (see Chapter 13). This rule worked so well that it had to be correct. But it was based on half-integral rather than integral spectral quantum numbers. Where did these half-integral values come from? In their attempts to answer this question, physicists were led inexorably to the discovery of electron spin.

There is a point we are trying to make here. Mathematics is the language of physics, and mathematical equations describe physical discoveries. But the mathematical equations sometimes become *physical pandoras*. Like the mythical Pandora's Box, they unleash consequences that were totally unsuspected by their creators.[1] Physics has quite a collection of *pandorae*, although they are not usually characterized in this manner. The *Landé interval rule* is a pandora. Invented to account for atomic spectra, it also gave us the spinning electron. The *Dirac equation* is another pandora. In order to satisfy the commutation requirements of a viable wave equation for the electron, Dirac was forced to use four-component spinors. However, only two components were needed for the two electron spin states. What did the other two components represent? As we now know, they represent the positron – the antiparticle of the electron. The Dirac equation led in a pandorean manner to the discovery of the positron. The *Maxwell equations* also turned out to be pandorean. The *divergences* and *curls* in these equations reproduced the conservation of electric charge and the twisting vector relationships of the magnetic field. But the Maxwell equations also revealed the close relationship between electromagnetic and optical phenomena, and thus bound these two disciplines forever together. Furthermore, the Maxwell equations and the constancy of the speed of light led Einstein to the theory of special relativity.

The *Planck radiation law* was at first regarded as just a mathematical device for solving a long-standing energy problem in black-body radiation. But Planck's quantized radiation field, after the work of Einstein and Compton, became recognized as the photon – the quantum of the electromagnetic field. The *pandorean pressure* of the Planck discovery, including his famous constant h, finally resulted in the development of quantum mechanics. Even *quantum electrodynamics* has turned out to be pandorean, in a story that is not yet complete. As formulated, *QED* was devised to account for the interactions between electrons and photons, and it accomplishes this with stunning accuracy. But it also accounts for the electromagnetic properties of muons, even though it has no explanation for the existence of the muon. And it doesn't really tell us very much about the electron itself, except for the important fact that its charge is point-like. As Feynman has noted,[2] *QED* gives the magnitude of the anomalous magnetic moment of the electron very accurately, but it doesn't tell us, except *via* direct calculation, whether the sign of this anomaly is positive or negative.

There is another strictly empirical result that may turn out to be *pandorean*. This is the systematics embodied in Figure 10.2, which shows the *thresholds* for producing $\mu\bar{\mu}$ muon pairs, $p\bar{p}$ proton pairs, $s\bar{s}$, $c\bar{c}$ and $b\bar{b}$ flavored-quark pairs, and the W and Z gauge bosons. As can be seen in this figure, the masses of these thresholds follow very accurate *mass triplings*, with the muon and proton grouped together in one family, and the $s\bar{s}$, $c\bar{c}$, $b\bar{b}$, W, and Z states grouped together in another. Now, the fact that the integral electric charge e seems to split apart into thirds in quark states suggests that mass triplings may be a natural phenomenon (Mac Gregor, 1990), with the mass multiplication somehow allied to the charge fractionization. But as Gottfried and Weisskopf (1984, p. 100), for example, have emphasized, there is no theory at present that explains triplings or any other kind of regularity in these mass values. What is even more strange about Figure 10.2 is that it shows the muon and proton grouped together in a mass-tripling family, even though they are thought to be totally unrelated types of particles. And it shows the flavored quarks and the gauge bosons grouped together in another mass-tripling family, even though these different types of resonances have only their unit spin values in common. The results displayed in Figure 10.2 seem too accurate and too comprehensive to be dismissed as accidental. Does the mass systematics that is contained in Figure 10.2 represent a modern-day Kepler ellipse, waiting for a modern-day Newton to come along? We have *mathematics* – the mass triplings – but no *physics*.

These examples lead us to what we might call the *Pandorean Principle of Physics*:

> *If a mathematical equation, or a set of mathematical equations, works so well in an experimental situation that it must be correct, then it is important to pursue all of its consequences, even those that seem unrelated to the original task at hand. The mathematical equations may contain more information about the world than the scientist who wrote them down.*

We have two applications of this *Pandorean Principle* that we would like to cite in finally concluding this discussion of *The Enigmatic Electron*. Both of these have to do with special relativity, as the title of this Postscript suggests. The first application is one that we have in fact already thoroughly discussed in Part III. This is the application of the Lorentz equations of special relativity to *rotating systems*, as embodied in the relativistically spinning sphere. The equations of special relativity were originally devised by Einstein to account for *(1)* the dynamics of *zero-mass* photons or electromagnetic waves, and *(2)* the dynamics of subluminally moving *massive* bodies such as the electron. The zero-mass objects move at the velocity of light, *c*, and thus represent a singular situation. The massive objects move at velocities less than *c*, and have total energies that vary in accordance with the Lorentz equations. These Lorentz equations are routinely applied today to linearly moving electrons (in an electron linac, for example), and to circularly moving electrons (in a betatron or microtron, for example). Thus they apply to both linear and curvilinear motion. However, they are *not* generally applied to the internal motion (the rotation) of the electron itself, although Einstein (1923, pp. 115-116) considered transformations between stationary and rotating frames of reference in some detail. The reason for this situation is of course the fact that unless the electron is very large, with a radius comparable to its Compton radius, then a real rotation that corresponds to its spin angular momentum of $1/2\ \hbar$ would have to involve velocities that exceed the limiting velocity *c*. Since these superluminal velocities are not permitted by the Lorentz equations, and since the belief in a point-like electron does not allow for a Compton-sized electron radius, there has been no motivation to pursue this line of reasoning. However, we have discovered in the present studies that by allowing the size of the electron to be large, and by making a natural application of special relativity to the rotational motion of the electron, we have obtained a set of equations that uniquely and accurately tie together the main spectroscopic properties of the electron. Perhaps the mathematics is telling us something! The equations of special relativity, as an example of the Pandorean Principle of Physics, may contain more information than we have realized.

The second *pandorean* application of the equations of special relativity does not involve the electron, but an even more arcane object – the *electron wave*. The electron wave is more mysterious than the electron. This wave clearly *does* things. It steers the electron in a double slit experiment, and it quantizes the electron orbitals in an atom: no coherent traveling wave, no double-slit interference pattern; no coherent standing wave, no quantized orbits. But, as presently envisaged, the electron wave itself *isn't* anything. According to the conventional Born interpretation, the electron wave is a *purely mathematical construct.* The electron is the physically real entity.[3] Thus we have the paradoxical situation of a physical system, the electron wave, that *is* nothing, and yet produces observable effects. But what does this have to do with special relativity? There is a pandorean connection between *special relativity* and the *electron wave*, and it has two things in common with the pandorean connection that we discussed above between *special relativity* and the *spinning electron*: *(1)* these pandorean connections both follow from simple and straightforward applications of the Lorentz equations; *(2)* the results they give are outside of the commonly recognized body of knowledge in present-day physics. The relationship that can be established between special relativity and the electron wave is contained in the equations of *perturbative special relativity* (Mac Gregor, 1985b), which we now briefly describe.

We know empirically that the electron generates some kind of electron wave as it moves through the vacuum state. The nature of this wave is completely unknown, but its wavelength and velocity are accurately specified by the de Broglie wave equation $\lambda = h/p$ and the de Broglie phase velocity equation $\mathrm{vV} = c^2$, where λ is the wavelength of the wave, p is the momentum of the electron, v is the electron velocity, and V is the wave phase velocity. Now suppose that the generation of this electron wave is kinematic, and is in agreement with the equations of special relativity. What must its velocity be? We can answer this question by setting up a simple situation. Consider a moving electron that "emits" a "particle". The properties of this *emitted particle* are unspecified except for the fact that it is "real": it carries both energy and momentum, which are related to one another via the Lorentz equations. The rest mass of the electron is unchanged during this particle emission process, and the entire process is required to conserve energy and momentum. This is a well-specified problem, and its solution merely requires writing down the Lorentz equations and solving them (although this solution is not to be found in any of the textbooks on special relativity). From the solution, we discover that the "emitted particle" is required to move forward at a velocity that exceeds the velocity of light, c. Thus this particle is a *tachyon*, and hence does not possess a rest mass in the conventional sense (its rest mass is imaginary). But it *does* possess a well-defined energy and well-defined linear momentum. Now suppose that we go to the *perturbative* limit, in which the energy and momentum of the emitted particle are only 10^{-6} or less of the electron energy and momentum. In this perturbative approximation, the forward component of the velocity of the emitted particle goes to a well-defined limit: *the forward velocity of the particle is equal to the de Broglie phase velocity.* Thus if the electron actually creates a real wave (an ensemble of "emitted particles") as it plows through the vacuum state, then the forward velocity of this wave, in order to conserve energy and momentum, is required to equal the postulated de Broglie phase velocity. Furthermore, this emitted wave is accurately planar (Mac Gregor, 1985b). Is the mathematics again telling us something? Is Pandora pointing the way? We are not accustomed to thinking with any seriousness about objects such as tachyons that travel faster than c. But we are familiar with photons that carry both energy and momentum *at* the velocity c. The concept of a real tachyon does not seem to be any more far-fetched than the concept of a purely fictitious (mathematical) electron wave that influences the motion of the electron.

The topic of the electron wave is a huge Pandora's Box – a new can of worms – and must be dealt with elsewhere.[4] The problem of classical *vis à vis* quantum physics, which is the central problem we have been addressing, logically calls into question both the nature of the electron and the nature of the electron wave.[5] It would be nice if Nature has in fact placed both of these entities within reach of our classical comprehension.

Notes

[1] This is what Wigner has described as the "unreasonable effectiveness" of mathematics (Schroeder, 1991, p. xiv).

[2] See the quotation by Feynman (1961a, pp. 75-76) in Chapter 8.

[3] See the quotation by Rosenfeld (1973, p. 260) in Chapter 4.

[4] For a discussion of electron waves, see Mac Gregor (1988).

[5] Perhaps the primary reason we study elementary particles is to see if they reveal what the fundamental building blocks of the universe really are. In studying the properties of the *electron*, we have (if the ideas presented in this book are correct) delineated the existence of a "point electrical charge" that has zero mass and holds itself together. We have also defined a "mechanical mass" that is continuous, rigid and non-interacting. If we were to extend these studies to include the *positron*, and specifically the electron-positron annihilation process, we would presumably be confronted with the necessity for a "mechanical anti-mass". Suppose we were to extend these studies even further, so that we focused our classical viewpoint on the *photon*. A key feature of the photon is that it has a very large angular momentum ($\hbar$) in relation to its total mass or energy (its rest mass is zero). It turns out that we can in fact construct a classical model for the photon, but in order to do this we need to invoke a new building block – "negative mass". The photon appears classically as a rotating positive-mass, negative-mass pair, where these masses carry electrical charges of $-e$ and $+e$. With appropriate (equal) values for the masses and for the separation distance d between the masses, this rotating pair of masses reproduces the angular momentum $\hbar$ of the photon (as evidenced in its electromagnetic interactions), and the centrifugal force of the rotation is counterbalanced by the electrostatic attraction between the two charges. The total energy of the photon, $\hbar\omega$, turns out to be equal to the electrostatic energy of the two charges, e^2/d. The "negative mass" contributes to the angular momentum of the photon (albeit in a rather strange way), and it cancels out the "positive mass" contribution to the energy. As a final thought, can we use this classical approach to study the *electron wave*? The electron wave has one property that is reminiscent of the photon: namely, it has no detectable energy, but it is able to produce momentum changes (it steers the electron). If we seek to construct a "classical model" of the electron wave, we are probably forced to make the structural elements out of matching positive-mass, negative-mass pairs. The concept of "negative mass" is not as far-fetched as it may at first seem. A negative mass can be treated as a "hole state", and its properties can be delineated in a very classical manner. As we know, a solid-state "vacancy" has many of the aspects of a particle. But a "hole state" implies that there is a "hole" or a "vacancy" in *something*. What is this *something*? Thus the pandorean story goes on and on.

Appendix A

Papers on Mott Single Scattering from Atomic Nuclei

One of the main purposes of the present monograph is to bring out the fact that Mott scattering at keV energies represents a potential way of detecting finite-size effects in electrons and positrons. These finite-size effects were discussed in Chapters 16 and 17. In single-scattering experiments, they show up in the form of *helical channeling*, which enhances the forward scattering and reduces the wide-angle scattering. In double-scattering experiments, they show up in the form of *helical depolarization*, which reduces the observed Mott polarization due to an effective rotation of the scattering plane. The finite-size effects are expected to appear in Mott scattering when the impact parameters for the scattering are comparable to the Compton wavelength of the electron or positron, which occurs for bombarding energies in the multi-keV region (above a few keV). These finite-size effects should disappear at energies above 1 MeV, where interaction times become so short that the helical motion is unimportant, and where the scattering of an essentially non-rotating rigid particle becomes point-like in nature due to the Golden Rule of Rigid Body Scattering (Chapter 15). Thus the energy window for observing finite-size effects in Mott scattering extends from perhaps 10 keV to a few hundred keV.

As we discussed in Chapter 17, some existing experiments exhibit anomalies which are suggestive of finite-size effects, but the data in the keV energy region are too fragmentary to permit any conclusions to be drawn. Thus new experiments seem to be required. With this idea in mind, it seems useful to present a listing of the references the author has gathered on the relevant aspects of Mott scattering. This listing is not claimed to be complete, but it should at least serve as a useful guide. The listing includes references to theoretical work, together with all the experimental references the author could find at energies ranging from 10 keV to a few MeV, plus some representative experimental references from the large volume of work done at energies below 10 keV. In Appendix A we list the references that apply to Mott single-scattering (mainly elastic scattering) experiments, and in Appendix B we list the references that apply to Mott double-scattering (asymmetry) experiments.

A. Theoretical Papers

Baluja, K. L., Jain, A., Jones, H. W., Weatherford, C. A. and Karim, K. R. (1991) *J. Phys. B: At. Mol. Opt. Phys.* **24**, L93.

Optical model calculations were carried out for positron-Ne scattering that include the positronium inelastic channel, and comparisons were made to experiment.

Bartlett, J. H. and Watson, R. E. (1940) *Proc. Am. Acad. Art Sci.* **74**, 53.

Mott elastic scattering cross sections were calculated for Hg at a number of energies.

Bartlett, J. H. and Welton, T. A. (1941) *Phys. Rev.* **59**, 281.

Mott screening corrections were calculated for Hg at 100 and 230 keV.

Basu, D., Basu, M. and Ghosh, A. S. (1987) *Atomic Physics with Positrons,* J. W. Humberston and E. A. G. Armour (eds), Plenum Press, New York, p. 377.

Calculations were made of positronium formation from electron scattering on hydrogen at 250 eV.

Bühring, W. (1968) *Z. Physik* **212**, 61.

Screened Mott elastic scattering cross sections and asymmetry functions were calculated for electron scattering on Hg at 46, 79, 100 and 204 keV over the angular range 45° - 180°. The calculated cross section values are in good agreement with the matching experimental values of Kessler (1968), as shown in the latter paper.

Burge, R. E. and Smith, G. H. (1962) *Proc. Phys. Soc.* **79**, 673.

A review is given of the theories used for electron microscopy. The electron energies cover the range from 20 to 1000 keV, but the scattering angles used are less than 1 degree.

Byatt, W. J. (1956) *Phys. Rev.* **104**, 1298.

Analytical fits were made to Hartree potentials for nineteen neutral atoms. These fits define coulomb potentials for use in Mott screening corrections.

Byron, F. W., Jr. and Joachin, C. J. (1977) *J. Phys. B: Atom. Molec. Phys.* ***10***, 207.

The elastic scattering of electrons and positrons by atomic H and He is calculated in the eikonal Born series (EBS) method, and is compared to experimental results on H at energies up to 400 eV, and on He at energies up to 700 eV.

Callaway, J. (1980) *Advances in Physics* **29**, 771.

A discussion is given of the theories of electron scattering by atoms in the energy range up to 100 eV.

Callaway, J. and Unnikrishnan, K. (1989) *Phys. Rev.* **A40**, 1660.

Optical model calculations were carried out for elastic electron scattering from H at energies between 30 to 400 eV, and were compared to experiment.

Curr, R. M. (1955) *Proc. Phys. Soc.* **A68**, 155.

The Mott scattering calculations of McKinley and Feshbach (1948) were redone, and additional terms were added to make the calculations extend accurately to high-Z elements.

Czyzewski, Z., MacCallum, D. O., Romig, A. and Joy, D. C. (1990) *J. Appl. Phys.* **68**, 3066.

The basic Mott scattering equations are discussed, and an extension is made to the case of electron scattering in solids, where a "muffin-tin" potential is used. Various screening potentials are described.

Dalitz, R. H. (1951) *Proc. Roy. Soc.* **206**, 509.

The scattering of a Dirac electron by a Yukawa potential is calculated in the second Born approximation.

Das, J. N. and Konar, A. N. (1974) *Nuovo Cimento* **21A**, 289.

Differential cross sections were calculated for inner-shell ionization by relativistic electrons.

Dewangan, D. P. and Eichler, J. (1989) *J. Phys. B: At. Mol. Opt. Phys.* **22**, L493.

The T matrix approach is applied to coulomb scattering.

Doggett, J. A. and Spencer, L. V. (1956) *Phys. Rev.* **103**, 1597.

The Mott elastic scattering cross section for electrons and positrons was evaluated at energies from 50 keV to 10 MeV for the atomic nuclei Z = 6, 13, 29, 50, 82 and 92.

Felder, R. M. (1970) *Nucl. Sci. Eng.* **42**, 232.

The Mott scattering equations were extended to include the multiple scattering theory of Goudsmit and Saunderson.

Fink, M. and Yates, A. C. (1970) *Atomic Data* **1**, 385. Part I.

The Dirac equation was used to calculate Mott elastic scattering and polarization for selected elements, using incident electron energies ranging from 100 to 1500 eV.

Fink, M. and Ingram, J. (1972) *Atomic Data* **4**, 129. Part II.

These are the same calculations as described in Fink (1970), but for different elements and wider (backscattering) angles.

Ford, G. W. and Mullin, C. J. (1958) *Phys. Rev.* **110**, 520.

An analysis was made of the processes that lead to inelastic electron scattering from atoms.

Franco, V. and Iwinski, Z. (1982) *Phys. Rev.* **A25**, 1900.

First-order corrections were made to the Glauber approximation for electron-hydrogen and positron-hydrogen scattering, and were compared to data from 30 to 680 eV.

Ghosh, A. S., Mukherjee, M. and Basu, M. (1989) *AIP Conference Proceedings* **205**, 622.

Calculations are described of low energy positron scattering on hydrogen.

Gregory, D. and Fink, M. (1974) *Atomic Data and Nuclear Data Tables* **14**, 39. Part III.

Dirac equation calculations for electron scattering and polarization are given for an additional series of elements. See Fink (1970) and Fink (1972).

Gupta, G. P. and Mathur, K. C. (1979) *J. Phys. B: Atom. Molec. Phys.* **12**, 3071.

The elastic scattering of electrons and positrons from He at 100 and 200 eV was calculated and compared to experiment.

Gupta, S. C. and Rees, J. A. (1975) *J. Phys. B: Atom. Molec. Phys.* **8**, 1267.

The elastic scattering of 100 eV electrons on He, Ne and Ar is discussed.

Hasenburg, K., Bartschat, K., McEachran, R. P. and Stauffer, A. D. (1987) *J. Phys. B: At. Mol. Phys.* **20**, 5165.

Theoretical calculations were made for elastic electron and positron scattering from Xe at energies up to 200 eV.

Holzwarth, G. and Meister, H. J. (1964) *Nucl. Phys.* **59**, 56.

Screened Mott cross sections and asymmetry functions were calculated for Au and Hg over a range of energies from 0.2 to 290 keV.

Ichimura, S., Aratama, M. and Shimizu, R. (1980) *J. Appl. Phys.* **51**, 2853.

Monte Carlo calculations were made of 1 - 20 keV electron backscattering from Al, Cu, Ag and Au.

Inokuti, M. (1971) *Rev. Mod. Phys.* **43**, 297.

The Bethe theory of the inelastic scattering of fast electrons by atoms is reviewed.

Jain, A., Etemadi, B. and Karim, K. R. (1990) *Physica Scripta* **41**, 321.

Total (elastic plus inelastic) cross sections were calculated for electron scattering from Ar and Kr at energies of 10 - 6000 eV.

Joachain, C. J., Vanderpoortens, R., Winters, K. H. and Byron, F. W. (1977) *J. Phys. B: Atom. Molec. Phys.* **10**, 227.

Optical model calculations were carried out for electron and positron scattering from Ar at energies of 100 to 1000 eV.

Joachain, C. J. (1989) *AIP Conference Proceedings* **205**, 68.

A discussion is given of recent advances in the theory of electron-atom collisions.

Kessler, J. (1976) *Polarized Electrons*, Springer-Verlag, Heidelberg.

This book gives a clear development of the Dirac equations as applied to Mott scattering and polarization experiments.

Khare, S. P., Kumar, A. and Lata, K. (1983) *J. Phys. B: Atom. Mol. Phys.* **16**, 4419.

Optical model calculations from 10 to 500 eV were made of total and differential elastic cross sections for electron and positron scattering from Mg.

Khare, S. P., Kumar, A. and Vijayshri (1985a) *J. Phys. B: Atom. Mol. Phys.* **18**, 1827.

Optical model calculations of electron and positron scattering on Ca were carried out at energies up to 500 eV for electrons and 75 eV for positrons.

Khare, S. P. and Lata, K. (1985b) *J. Phys. B: At. Mol. Phys.* **18**, 2941.

Fredholm integrals were used to calculate the elastic scattering of electrons and positrons on He and H_2 at energies up to 1 keV.

Khare, S. P., Kumar, A. and Lata, K. (1986) *Phys. Rev.* **A33**, 2795.

An optical model calculation was made of positron scattering from Ar at energies between 100 and 800 eV.

Lin, S., Sherman, N. and Percus, J. K. (1963) *Nucl. Phys.* **45**, 492.

Screened Mott elastic scattering cross sections and asymmetry functions were calculated for Au at 121 keV and for Hg at 46 and 79 keV, over a range of scattering angles from 15° to 165°.

Lin, S. (1964) *Phys. Rev.* **133**, A965.

Screened Mott elastic scattering cross sections and asymmetry functions were calculated for Au and Cu at energies from 50 to 400 keV, and for Hg at 204 keV, over a range of scattering angles from 10° to 170°.

Madison, D. H. (1989) *AIP Conference Proceedings* **205**, 149.

Theoretical calculations were made of elastic and inelastic electron scattering from hydrogen.

Mathur, K. C. and Purohit, S. P. (1990) *Phys. Rev.* **A42**, 2696.

Theoretical calculations were made of the elastic scattering of electrons and positrons by Li at energies up to 60 eV, and comparisons were made to the available experimental data.

McEachran, R. P. and Stauffer, A. D. (1984) *J. Phys. B: At. Mol. Phys.* **17**, 2507.

Calculations were made for the elastic scattering of electrons on Kr and Xe at energies up to 50 eV.

McKinley, W. A. and Feshbach, H. (1948) *Phys. Rev.* **74**, 1759.

The Mott scattering equations were evaluated and extended to include some higher-order terms.

Meister, H. J. and Weiss, H. F. (1968) *Z. Physik* **216**, 165.

The non-relativistic (Schrödinger) and relativistic (Dirac) electron scattering equations were compared at low energies (below 200 eV) in order to assess the importance of relativistic effects on differential cross sections and spin polarizations.

Mohr, C. B. O. and Tassie, L. J. (1954) *Proc. Phys. Soc.* **A67**, 711.

Screened Mott elastic scattering cross sections and asymmetry functions were calculated for Au over an energy range from 0.9 to 121 keV.

Moore, P. and Fink, M. (1972) *Phys. Rev.* **A5**, 1747.

Relativistic calculations of elastic differential cross sections for electron scattering on Hg at 46, 79, and 100 keV were carried out using a Liberman potential for the screening corrections.

Mott, N. F. and Massey, H. S. W. (1949) *The Theory of Atomic Collisions*, Second edition, Clarendon Press, Oxford.

Pages 74-87 of this reference give the basic Dirac equations for electron and positron Mott scattering, and screening corrections are discussed on pages 118 and 195-196.

Motz, J. W., Olson, H. and Koch, H. W. (1964) *Rev. Mod. Phys.* **36**, 881.

This is an extensive review of the Mott-type calculations that have been made with respect to the elastic scattering and spin polarization of electrons by impacts with atomic nuclei. Screening effects are discussed in detail. Numerical values for calculations by various authors are given.

Mukherjee, M., Basu, M. and Ghosh, A. S. (1987) *Phys. Rev.* **A36**, 937.

Elastic electron-H scattering was calculated from 30 - 680 eV and compared to experiment.

Nahar, S. N. and Wadehra, J. M. (1991) *Phys. Rev.* **A43**, 1275.

Calculations of Mott scattering cross sections and P, T and U polarization vectors were carried out for electron and positron scattering off Ar at energies up to 300 eV, using a detailed treatment of the Dirac equation. Comparisons were made to experiment.

Naon, M. and Cornille, M. (1972) *J. Phys. B: Atom. Molec. Phys.* **5**, 1965.

A theoretical formalism is developed that uses small-angle keV electron scattering to deduce electronic correlation functions for Ne.

Naon, M., Cornille, M. and Kim, Y. (1975) *J. Phys. B: Atom. Molec. Phys.* **8**, 864.

The work of Naon (1972) is extended to include Ar.

Nigam, B. P., Sundaresan, M. K. and Wu, T.-Y. (1959) *Phys. Rev.* **115**, 491.

Multiple scattering theory, for the case of the scattering of fast electrons by atoms, was extended to include the second Born approximation, and was used to correct the corresponding Molière expression.

Peixoto, E. M. A., Bunge, C. F. and Bonham, R. A. (1969) *Phys. Rev.* **181**, 322.

An investigation was made of atomic electron correlation functions as measured by elastic and inelastic electron scattering.

Riley, M. E., MacCallum, C. J. and Biggs, F. (1975) *Atomic Data and Nuclear Data Tables* **15**, 443.

Calculations of theoretical electron-atom elastic scattering cross sections were made for twenty four elements over the energy range from 1 to 256 keV.

Rose, M. E. (1940) *Phys. Rev.* **57**, 280.

A review is given of the basic Mott equations for the scattering and polarization of electrons, and the experimental results up to that time (which were not very accurate) are discussed.

Rudge, M. R. H. (1968) *Rev. Mod. Phys.* **40**, 564.

A review is given of the quantum theory of the ionization of atoms by electron impact.

Sarkar, K. P. and Ghosh, A. S. (1989) *J. Phys. B: At. Mol. Opt. Phys.* **22**, 105.

Elastic differential cross sections for positron scattering on Li and Na were calculated for energies up to 100 eV.

Scholz, T., Scott, P. and Burke, P. G. (1988) *J. Phys. B: At. Mol. Opt. Phys.* **21**, L139.

R-matrix theory was used to calculated electron-H scattering.

Scholz, T. T., Walters, H. R. J. and Burke, P. G. (1990) *J. Phys. B: At. Mol. Opt. Phys.* **23**, L467.

Calculations were made of total electron-H cross sections up to a few eV.

Scholz, T. T., Walters, H. R. J., Burke, P. G. and Scott, M. P. (1991) *J. Phys. B: At. Mol. Opt. Phys.* **24**, 2097.

Calculations were made of differential elastic and inelastic electron-H cross sections up to a few eV.

Scott, M. P., Scholz, T. T., Walters, H. R. J. and Burke, P. G. (1989) *J. Phys. B: At. Mol. Opt. Phys.* **22**, 3055.

Calculations were made of elastic and inelastic electron-H cross sections for energies up to a few eV.

Scott, W. T. (1963) *Rev. Mod. Phys.* **35**, 231.

A review is given of multiple scattering theory as applied to the small-angle scattering of fast charged particles.

Seaton, M. J. (1969) *Rivista del Nuovo Cimento* **1**, 391.

A review is given of elastic and inelastic electron-atom collision theory.

Sherman, N. (1956) *Phys. Rev.* **103**, 1601.

Unscreened Mott cross sections and asymmetry functions were calculated from the Dirac equation for Al, Cd and Hg at a variety of electron energies.

Sherman, N. and Nelson, D. F. (1959) *Phys. Rev.* **114**, 1541.

Unscreened Mott cross sections and asymmetry functions were calculated for Au and Al at 75 and 121 keV.

Stauffer, A. D. *et al.* (1989) *AIP Conference Proceedings* **205**, 645.

A discussion is given of theoretical calculations of positron scattering from alkali atoms and noble gases, and comparisons are made with experimental data.

Tilinin, I. S. (1988) *Zh. Eksp. Teor. Fiz.* **94**, 96.

Transport cross sections were calculated for electron and positron scattering from complex atoms.

Walker, D. W. (1971) *Advances in Physics* **20**, 257.

The use of the Dirac equation in calculating low energy electron elastic scattering and spin polarization from atoms is discussed in detail.

Walters, H. R. J. (1984) *Physics Reports* **116**, 1.

A review is given of perturbative methods used to calculate electron and positron scattering from atoms, and extensive comparisons of experimental electron scattering data with theory are presented.

Walters, H. R. J. (1987) *Atomic Physics with Positrons,* J. W. Humberston and E. A. G. Armour (eds), Plenum Press, New York, p. 449.

Calculations of positron scattering from atomic hydrogen scattering at energies between 54 and 300 eV are described.

Weber, T. A., Deck, R. T. and Mullin, C. J. (1963) *Phys. Rev.* **130**, 660.

Calculations were made of low-energy electron spectra resulting from the scattering of high-energy electrons on bound atomic electrons.

Zeitler, E. and Olsen, H. (1964) *Phys. Rev.* **136**, A1546.

Various screening approximations for Mott elastic scattering were investigated. Results are shown for Cu and Au at 50, 100, 200 and 400 keV.

Zeitler, E. and Olsen, H. (1966) *Z. Naturforsch.* **21a**, 1321.

Calculations of screening corrections to Mott elastic scattering cross sections were made for C, Al, Cu, Sn, Pb and U at 0.2, 0.4, 0.7, 1, 2, 4 and 10 MeV.

Zeitler, E. and Olsen, H. (1967) *Phys. Rev.* **162**, 1439.

A discussion is given of complex scattering amplitudes that are important in Mott elastic scattering screening calculations,

B. Experimental Papers Below 10 keV

Bransden, B. H. and McDowell, M. R. C. (1978) *Physics Reports* **46**, 249.

A detailed (146 page) summary is given of theoretical and experimental electron scattering by light atoms at energies up to about one keV.

Bromberg, J. P. (1969) *J. Chem. Phys.* **51**, 4117.

Absolute elastic differential cross sections were measured for electron scattering on Hg at 500, 400 and 300 eV.

Bromberg, J. P. (1974) *J. Chem. Phys.* **61**, 963.

Absolute differential cross sections were measured for electron scattering on He, Ne, Ar, Kr, and Xe at energies of 200, 300, 400, 500, and 700 eV, and at scattering angles from 2° to 30°.

Bronshtein, I. M. and Pronin, V. P. (1976a) *Sov. Phys. Solid State* **17**, 1363.

Measurements were made of the angular distributions of elastically backscattered electrons from Be, Al, Cu, Ag, and Au, using 0.1 to 2 keV electrons incident perpendicular to the metal surface.

Bronshtein, I. M. and Pronin, V. P. (1976b) *Sov. Phys. Solid State* **17**, 1610.

Measurements were made of the angular distributions of elastically backscattered electrons from Be, Al, Cu, Ag, and Au, using 0.1 to 2 keV electrons incident at an angle of 75° to the metal surface.

Bronshtein, I. M. and Pronin, V. P. (1976c) *Sov. Phys. Solid State* **17**, 1672.

A study was made of the elastic backscattering of electrons from composite Be and Au targets.

Buckman, S. J., Noble, C. J. and Teubner, P. J. O. (1979) J. Phys. B: Atom. Molec. Phys. **12**, 3077.

Differential cross sections were measured for the elastic and inelastic scattering of electrons on K up to 200 eV, and comparisons were made to optical model calculations.

Charlton, M. (1985) *Rep. Prog. Phys.* 48, 737.

A review is given of experimental studies of positron scattering in gases.

Coleman, P. G. and McNutt, J. D. (1979) *Phys. Rev. Lett.* **42**, 1130.

Differential elastic cross sections for positron scattering on Ar were measured at energies between 2.2 and 8.7 eV.

Danjo, A. (1988) *J. Phys. B: At. Mol. Opt. Phys.* **21**, 3759.

Elastic differential cross sections for electron scattering from Kr were measured at energies from 5 to 200 eV.

De Heer, F. J, Jansen, R. H. J. and van der Kaay, W. (1979) *J. Phys. B: Atom. Molec. Phys.* **12**, 979.

Total cross sections for electron scattering on Ne, Ar, Kr and Xe were deduced from experimental and theoretical data on elastic and inelastic scattering, including some elastic and inelastic differential cross sections. The energy range covered was 20 eV to 3 keV.

DuBois, R. D. and Rudd, M. E. (1976) *J. Phys. B: Atom. Molec. Phys.* **9**, 2657.

Absolute differential cross sections were measured for the elastic scattering of electrons up to 800 eV from Ar, Ne, N_2 and CO_2.

Ebel, F. *et al.* (1987) *Atomic Physics with Positrons*, J. W. Humberston and E. A. G. Armour (eds), Plenum Press, New York, p. 391.

The generation of slow (300 eV) positrons by the 65 MeV Giessen linac is described.

Fromme, D., Kruse, G., Raith, W. and Sinapius, G. (1987) *Atomic Physics with Positrons*, J. W. Humberston and E. A. G. Armour (eds), Plenum Press, New York, p. 403.

Comparisons are made between positron and proton scattering from helium.

García, G., Arqueros, F. and Campos, J. (1986) *J. Phys. B: At. Mol. Phys.* **19**, 3777.

Total cross sections were measured for electron scattering from Ne, Ar and Kr at energies of 700 to 6000 eV.

García, G., Pérez, A. and Campos, J. (1988) *Phys. Rev.* **A38**, 654.

Total cross sections were measured for electron scattering from N_2 molecules at energies of 600 to 5000 eV.

García, G., Aragón, C. and Campos, J. (1990) *Phys. Rev.* **A42**, 4400.

Total cross sections were measured for electron scattering from CO molecules at energies of 380 to 5200 eV.

Gergely, G. (1987) *Vacuum* **37**, 149.

Angular distributions of elastically backscattered electrons off Al, Fe, V, Cu, Ag, Ta, W and Au were measured at electron energies up to 2 keV.

Holtkamp, G., Jost, K., Peitzmann, F. J. and Kessler, J. (1987) *J. Phys. B: At. Mol. Phys.* **20**, 4543.

Absolute differential cross sections were measured for elastic electron scattering off Hg atoms at energies of 25 to 300 eV.

Howell, R. H., Fluss, M. J., Rosenberg, I. J. and Meyer, P. (1985) *Nucl. Instr. Meth.* **B10/11**, 373.

A description is given of the LLNL electron linac positron source, which can produce positrons with energies between 500 eV and 20 keV.

Howell, R. H. (1987) *Atomic Physics with Positrons*, J. W. Humberston and E. A. G. Armour (eds), Plenum Press, New York, p. 215.

Positronium formation from positron bombardment of solids was studied for positron energies up to 3 keV.

Humberston, J. W. (1988) *Comments on Atomic and Molecular Physics* **21**, 333.

A discussion is given of the status of low energy positron experiments.

Hyder, G. M. A. *et al.* (1986) *Phys. Rev. Lett.* **57**, 2252.

Relative differential elastic cross sections were measured in a crossed-beam experiment for positron and electron scattering from argon at energies of 100 to 300 eV.

Iga, I., Lee, M.-T., Nogueira, J. C. and Barbieri, R. S. (1987) *J. Phys. B: At. Mol. Phys.* **20**, 1095.

Electron elastic differential cross sections were measured for scattering off Ar and O_2 at energies from 300 to 1000 eV, and were compared to theoretical calculations.

Jansen, R. H. J., de Heer, F. J., Luyken, H. J. and van Wingerden, B. (1976a) *J. Phys. B: Atom. Molec. Phys.* 9, 185.

Absolute differential cross sections were measured for the elastic scattering of electrons by He, Ne, Ar and N_2 at energies up to 3 keV.

Jansen, R. H. J. and de Heer, F. J. (1976b) *J. Phys. B: Atom. Molec. Phys.* **9**, 213.

This is an extension of Jansen (1976a) to include Kr and Ar.

Joachain, C. J. (1987) *Atomic Physics with Positrons,* J. W. Humberston and E. A. G. Armour (eds), Plenum Press, New York, p. 71.

A comprehensive discussion is given of low energy positron experiments on atoms.

Jost, K. and Ohnemus, B. (1979) *Phys. Rev.* **A19**, 641.

Absolute total cross sections for electron-mercury scattering were measured at energies between 0.1 and 500 eV.

Kauppila, W. E. and Stein, T. S. (1987) *Atomic Physics with Positrons,* J. W. Humberston and E. A. G. Armour (eds), Plenum Press, New York, p. 27.

Positron differential cross section measurements on Ar at low energies are described.

Kauppila, W. E. and Stein, T. S. (1989) *AIP Conference Proceedings* **205**, 627.

A discussion is given of measurements of total cross sections on Na and K and differential elastic cross sections on the inert gases.

Kessler, J. and Lindner, H. (1965) *Z. Physik* **183**, 1.

Angular distribution measurements were made of low-energy (200 eV to 4000 eV) electron scattering off mercury atoms. The angular distributions exhibited characteristic diffraction peaks.

Kieffer, L. J. and Dunn, G. H. (1966) *Rev. Mod. Phys.* **38**, 1.

This is a survey of the experimental data on low energy electron impact ionization cross sections.

Klewer, M., Beerlage, M. J. M. and van der Wiel, M. J. (1980) *J. Phys. B: Atom. Molec. Phys.* **13**, 571.

In an electron-Xe crossed-beam experiment, the angular distributions of electrons elastically scattered from Xe were measured at energies between 2 and 300 eV.

Lesiak, B., Jablonski, A. and Gergely, G. (1990) *Vacuum* **40**, 67.

Elastic peak electron spectroscopy was used to determine electron inelastic mean free paths from backscattering studies (see Gergely, 1987).

Lubell, M. S. (1987) *Atomic Physics with Positrons*, J. W. Humberston and E. A. G. Armour (eds), Plenum Press, New York, p. 287.

A description is given of the BNL positron-H scattering experiment, which features a high-intensity ^{22}Na reactor positron source with an energy range of 10 to 2000 eV.

Lynn, K. G., Weber, M., Roellig, L. O., Mills, A. P., Jr. and Moodenbaugh, A. R. (1987) *Atomic Physics with Positrons*, J. W. Humberston and E. A. G. Armour (eds), Plenum Press, New York, p. 161.

A description is given of a ^{64}Cu low energy positron source obtained from the BNL high thermal neutron flux reactor.

Marinkovic, B., Pejcev, V., Filipovic, D. and Vuskovic, L. (1991) *J. Phys. B: At. Mol. Opt. Phys.* **24**, 1817.

Elastic and inelastic electron scattering cross sections were measured for Cd at energies up to 85 eV.

McGuire, J. H. and Deb, N. C. (1987) *Atomic Physics with Positrons*, J. W. Humberston and E. A. G. Armour (eds), Plenum Press, New York, p. 83.

A discussion is given of possible positron experiments in atomic physics.

Mehr, J. (1967) *Z. Physik* **198**, 345.

Elastic scattering angular distributions were measured for 5 to 1000 eV electrons incident on He, Ne, Ar, Kr, and Xe, and the spin polarization angular distribution was measured for 40 eV electrons incident on Ar.

Mitroy, J., McCarthy, I. E. and Stelbovics, A. T. (1987) *J. Phys. B: At. Mol. Phys.* **20**, 4827.

Theoretical and experimental studies were made of electron scattering from Na at energies above 10 eV.

Moiseiwitsch, B. L. and Smith, S. J. (1968) *Rev. Mod. Phys.* **40**, 238.

An extensive review is given of the electron impact excitation of atoms.

Nogiera, J. C., Iga, I. and Lee, M.-T. (1982) *J. Phys. B: At. Mol. Phys.* **15**, 2539.

Absolute total cross sections for electron scattering on Ar were measured at energies between 0.5 and 3 keV.

Peitzmann, F. J. and Kessler, J. (1990) *J. Phys. B: At. Mol. Opt. Phys.* **23**, 4005.
A check was made of absolute differential cross section measurements on Hg at energies up to 300 eV, using a crossed-beam technique.

Raith, W. and Sinapius, G. (1989a) *Comments At. Mol. Phys.* **22**, 199.
A discussion is given as to what can be learned about electron-atom collisions by studying positron-atom collisions.

Raith, W., Olsson, B., Sinapius, G., Sperber, W. and Spicher, G. (1989b) *AIP Conference Proceedings* **205**, 633.
The electron and positron impact ionization of H was measured in a crossed-beam experiment over the energy range 30 to 600 eV.

Rao, M. V. V. S. and Bharathi, S. M. (1987) *J. Phys. B: At. Mol. Phys.* **20**, 1081.
Relative differential cross section measurements were made for the elastic scattering of 150 to 500 eV electrons from Na, K and Rb.

Read, F. H. (1983) *Physica Scripta* **27**, 103.
A review is given of the status of electron-atom scattering experiments at low energies.

Register, D. F., Vuskovic, L. and Trajmar, S. (1986) *J. Phys. B: At. Mol. Phys.* **16**, 1685.
Elastic Xe differential cross sections were measured for electron scattering at energies from 1 to 100 eV.

Rubin, K. and Efremov, I. (1985) *Phys. Rev. Lett.* **54**, 1150.
Experimental anomalies were found in 4 eV electron-potassium scattering at extreme forward angles.

Saha, B. C., Chaudhuri, J. and Ghosh, A. S. (1976) *J. Phys. Soc. Japan* **41**, 1716.
Electron-hydrogen elastic scattering at 100 - 200 eV was calculated and compared to experiment.

Schwab, A., Höner, P., Raith, W. and Sinapius, G. (1987) *Atomic Physics with Positrons*, J. W. Humberston and E. A. G. Armour (eds), Plenum Press, New York, p. 429.
A crossed-beam positron experiment on Ar was carried out at 30 eV.

Sinapius, G. and Ravn, H. L. (1987) *Atomic Physics with Positrons*, J. W. Humberston and E. A. G. Armour (eds), Plenum Press, New York, p. 185.
A discussion is given of short-lived positron radioisotope sources.

Srivastava, S. K. and Vuskovic, L. (1980) *J. Phys. B: Atom. Molec. Phys.* **13**, 2633.
A crossed-beam technique was used to measure elastic and inelastic electron-Na scattering cross sections from 10 to 54 eV.

Stein, T. S., Gomez, R. D., Hsieh, Y.-F., Kauppila, W. E., Kwan, C. K. and Wan, Y. J. (1985) *Phys. Rev. Lett.* **55**, 488.
Total cross sections were measured between 5 and 49 eV for electrons and positrons colliding with K atoms.

Stein, T. S., Dababneh, M. S., Kauppila, W. E., Kwan, C. K. and Wan, Y. J. (1987) *Atomic Physics with Positrons,* J. W. Humberston and E. A. G. Armour (eds), Plenum Press, New York, p. 251.
Positron and electron total scattering measurements on alkali atoms are described.

Van Wyngaarden, W. L. and Walters, H. R. J. (1986) *J. Phys. B: At. Mol. Phys.* **19**, 929.
Elastic and inelastic electron-hydrogen atomic scattering calculations from 54.4 to 350 eV were carried out and compared to experiment.

Vuskovic, L., Maleki, L. and Trajmar, S. (1984) *J. Phys. B: At. Mol. Phys.* **17**, 2519.
Elastic and inelastic differential cross sections were measured for electron scattering on Rb atoms from 10 to 200 eV.

Wadehra, J. M. and Nahar, S. N. (1987) *Atomic Physics with Positrons,* J. W. Humberston and E. A. G. Armour (eds), Plenum Press, New York, p. 445.
Calculations of positron-Ar scattering from 3 to 300 eV are described.

Wagenaar, R. W. and de Heer, F. J. (1980) *J. Phys. B: Atom. Molec. Phys.* **13**, 3855.
Total cross sections for electron scattering from Ne, Ar, Kr and Xe were measured at energies between 25 and 750 eV.

Wagenaar, R. W., de Boer, A., van Tubergen, T., Los, J. and de Heer, F. J. (1986) *J. Phys. B: At. Mol. Phys.* **19**, 3121.
Absolute differential cross sections for small-angle elastic electron scattering from noble gases were measured at energies between 20 and 200 eV.

Williams, J. F. and Willis, B. A. (1975a) *J. Phys. B: Atom. Molec. Phys.* **8**, 1641.
Differential cross sections for inelastic electron scattering from H were measured from 54 to 680 eV.

Williams, J. F. and Willis, B. A. (1975b) *J. Phys. B: Atom. Molec. Phys.* **8**, 1670.
Absolute differential elastic cross sections were measured for 20 to 400 eV electron scattering from argon.

Williams, J. F. (1975c) *J. Phys. B: Atom. Molec. Phys.* **8**, 1683.

Absolute differential elastic cross sections were measured for 0.5 to 8.7 eV electron scattering from atomic hydrogen.

Williams, J. F. (1975d) *J. Phys. B: Atom. Molec. Phys.* **8**, 2191.

Absolute differential elastic cross sections were measured for electron scattering from atomic hydrogen at energies of 20 to 680 eV.

Williams, J. F. and Crowe, A. (1975e) *J. Phys. B: Atom. Molec. Phys.* **8**, 2233.

Absolute differential cross sections were measured for electron scattering from Ne, Kr and Xe at energies of 20 to 400 eV.

Williams, J. F. and Allen, L. J. (1989) *J. Phys. B: At. Mol. Opt. Phys.* **22**, 3529.

Electron-oxygen elastic cross sections were measured at energies from 0.5 to 8.7 eV.

Williams, W., Trajmar, S. and Bozinis, D. (1976) *J. Phys. B: Atom. Molec. Phys.* **9**, 1529.

Elastic and inelastic electron scattering from Li was measured at energies between 5.4 and 60 eV.

Zecca, A., Grzegorz, K., Brusa, R. S. and Grisenti, R. (1991) *J. Phys. B: At. Mol. Opt. Phys.* **24**, 2737.

Measurements of absolute total cross sections were made for electron scattering from Kr and Xe over an energy range of 80 to 4000 eV.

C. Experimental Papers Above 10 keV

Anthony, D. J., Waldman, B. and Miller, W. C. (1953) *Phys. Rev.* **91**, 439a.

This A. P. S. abstract describes Mott elastic electron scattering from thin Al and Au foils at an energy of 1700 keV.

Barber, A. and Champion, F. C. (1938) *Proc. Roy. Soc.* **A168**, 159.

This is early work, in which RaE beta rays with energies from 490 to 1125 keV were scattered off Hg atoms.

Bargholtz, Chr. *et al.* (1987) *J. Phys. G: Nucl. Phys.* **13**, L265.

Bhabha positron scattering from Th was measured at energies between 265 and 425 keV. This was an "axion" search, and no Mott scattering was observed.

Bayard, R. T. and Yntema, J. L. (1955) *Phys. Rev.* **97**, 372.

Relative differential cross sections for Mott electron scattering off Al and Au nuclei were measured at 600, 1000, and 1700 keV, over an angular range from 30 to 150 degrees. Absolute 60 degree cross sections were measured for Al at these energies.

Braicovich, L., De Michelis, B. and Fasana, A. (1967) *Phys. Rev.* **154**, 234.

The plural scattering of electrons and positrons from thin gold foils was investigated for incident energies between 180 and 395 keV.

Brown, B., Matsukawa, E. and Stewardson, E. A. (1956) *Proc. Phys. Soc.* **69**, 496.

Measurements were made of the Mott scattering of 400 and 500 keV electrons from U foils on Al backings, using scattering angles of 90° and 105°. Measurements were also made for pure Al foils at these same energies and angles.

Buechner, W. W., Van de Graaff, R. J., Sperduto, A., Burrill, E. A., Jr. and Feshbach, H. (1947) *Phys. Rev.* **72**, 678.

Measurements were made of the Mott scattering of 2100, 2200 and 2300 keV electrons from Be and Al, using scattering angles from 35 to 60 degrees.

Champion, F. C. and Barber, A. (1939) *Phys. Rev.* **55**, 111.

The elastic scattering of 700 to 1200 keV electrons off iodine nuclei was studied in a cloud chamber.

Chapman, K. R., Matsukawa, E., Rose, P. H. and Stewardson, E. A. (1955) *Proc. Phys. Soc.* **68**, 928.

Absolute measurements were made of back-angle Mott scattering off an aluminum foil at electron energies of 400 and 500 keV. The counter was a galvanometer with an absorber in front to remove scattered electrons.

Chase, C. T. and Cox, R. T. (1940) *Phys. Rev.* **58**, 243.

Measurements were made of 50 keV electrons scattered off an aluminum foil.

Cooper, J. W. and Kolbenstvedt, H. (1972) *Phys. Rev.* **A5**, 677.

Calculations were made of 100 - 400 keV differential cross sections for the electron ionization of C, Cu and Au, and comparisons were made with recent experimental results.

Cosslett, V. E. and Thomas, R. N. (1964a) *Brit. J. Appl. Phys.* **15**, 883.

Measurements were made of multiple scattering effects for 5 - 30 keV electrons incident on thin evaporated films of Al, Cu, Ag and Au.

Cosslett, V. E. and Thomas, R. N. (1964b) *Brit. J. Appl. Phys.* **15**, 1283.

The measurements cited in Cosslett (1964a) were used to derive range-energy relations for electron scattering in Al, Cu, Ag and Au.

Cosslett, V. E. and Thomas, R. N. (1965) *Brit. J. Appl. Phys.* **16**, 779.

The measurements cited in Cosslett (1964a) were used to deduce backscattering effects.

Cox, R. T. and Chase, C. T. (1937) *Phys. Rev.* **51**, 141.
Measurements were made of 80 keV electrons scattering off a gold foil.

Damodaran, K. K. and Curr, R. M. (1956) *Proc. Phys. Soc.* **69**, 196.
Relative measurements were made of the differential elastic scattering of 4330 keV electrons off Al, Ag, Pt and U targets, using emulsion detectors. Absolute measurements were made for Al and U.

Dick, C. E. and Motz, J. W. (1968) *Phys. Rev.* **171**, 75.
Inelastic scattering cross sections were measured for 200 and 400 keV electrons incident on Al, Cu, Sn and Au.

Dougal, R. C. (1965) *Proc. Phys. Soc.* **85**, 851.
Mott scattering was measured on gold over the energy range 100 - 400 keV, and was expressed as ratios of the scattering at angles of 20°, 30° and 45° to the scattering at 65°. The measurements demonstrated the importance of screening corrections.

Duguet, A., Lahmam Bennani, A. and Roualt, M. (1983) *J. Chem. Phys.* **79**, 2786.
Relative measurements were made of forward angle (up to 13°) elastic and inelastic scattering angular distributions for electrons on Ne and Ar at 25 keV.

Erman, P. (1970) *Physica Scripta* **1**, 93.
Electrons and positrons from 100 to 1300 keV, obtained by using radioactive sources in a beta ray spectrometer, were used to investigate plural and multiple scattering in moderately thick foils of Al, Cu, Ag and Pb.

Fink, M. and Kessler, J. (1966) *Z. Physik* **196**, 1.
Absolute values of electron scattering cross sections on Ar, Kr and Xe at small forward scattering angles (up to 65 mrad) were measured for incident electron energies of 37, 60 and 100 keV.

Fink, M. and Bonham, R. A. (1969) *Phys. Rev.* **187**, 114.
Total differential (elastic plus inelastic) scattering cross sections over small forward angles were measured for 40 keV electrons incident on Ne.

Fink, M. and Moore, P. G. (1977) *Phys. Rev.* **A15**, 112.
Precision measurements of total differential (elastic plus inelastic) scattering cross sections from 7 to 200 mrad were made for 40 keV electrons incident on Ne and Ar.

Fink, M., Moore, P. G. and Gregory, D. (1979a) *J. Chem. Phys.* **71**, 5227.
Precision measurements of total differential (elastic plus inelastic) scattering cross sections at small angles (up to 200 mrad) were made for 30, 40 and 50 keV electrons incident on molecular N_2.

Fink, M., Schmiedekamp, C. W. and Gregory, D. (1979b) *J. Chem. Phys.* 71, 5238 (1979).

Precision measurements of total differential (elastic plus inelastic) scattering cross sections at small angles (up to 200 mrad) were made for 50 keV electrons incident on CH_4, CO_2 and CF_4.

Grachev, B. D., Kozlovskii, S. S., Korobochko, Yu. S., Mineev, V. I. and Petrochenko, A. F. (1978) *Sov. Tech. Phys. Lett.* **4**, 110.

Relative elastic and inelastic electron scattering cross sections off a thin copper film were measured at 98.6 keV.

Grachev, B. D., Kozlovskii, S. S., Korobochko, Yu. S., Mineev, V. I. and Petrochenko, A. F. (1980) *Sov. Phys. (JETP)* **52**, 827.

This work is a continuation of the studies described in Grachev (1978).

Hanson, A. O. (1987) *AIP Conference Proceedings* **161**, 12.

This is a discussion of the earliest work on Mott scattering, plus a description of the 15.7 MeV Illinois electron scattering experiments (Lyman, 1951) that showed a finite nuclear size.

Kanter, H. (1964) *Brit. J. Appl. Phys.* **15**, 555.

The backscattering of electrons from thin films was measured, using 1 to 15 keV electrons incident on thin Al and Au films.

Keck, K. (1962) *Z. Physik* **167**, 468.

Measurements were made of the scattering of 70 keV electrons from collodion foils embedded with lead.

Kessler, J. (1959) *Z. Physik* **155**, 350.

Relative cross sections were measured for the scattering of 150 keV electrons off gas targets of Kr, Xe, Hg, using an annular target geometry that subtended scattering angles of 60, 90, 120 and 135 degrees.

Kessler, J. and Weichert, N. (1968) *Z. Physik* **212**, 48.

Electrons of 46, 79, 100 and 204 keV were scattered by a mercury atomic beam, and were measured at scattering angles from 45° to 135°, in an experiment designed to study the effects of screening. The beam was calibrated by measuring the scattering from a comparison hexadecane ($C_{16}H_{34}$) beam. The experimental cross sections agree well with matching theoretical screened cross sections calculated by Bühring (1968) and by Moore and Fink (1972).

Ketkar, S. N., Fink, M. and Bonham, R. A. (1983) *Phys. Rev.* **A27**, 806.

Total (elastic plus inelastic) differential cross sections at forward angles were measured to an accuracy of 0.1% for electron scattering on He at 40 keV .

Kinzinger, E. and Bothe, W. (1952) *Z. Naturforschg.* **7A**, 390.

The electron scattering cross sections of Ni, Ag and Au relative to Al were measured at 245 keV, and absolute cross sections for Al were measured at this same energy.

Kinzinger, E. (1953) *Z. Naturforschg.* **8A**, 312.

Electron scattering from Al and Au was measured from 150 to 400 keV at a scattering angle of 120°, and complete angular distributions for Al and Au were measured at an energy of 245 keV.

Klarmann, H. and Bothe, W. (1936) *Z. Physik* **101**, 489.

Electron scattering from Xe and Kr was measured.

Kobayashi, T. and Shimizu, S. (1972) *J. Phys. B: Atom. Molec. Phys.* **5**, L211.

Differential cross sections for the elastic scattering of positrons by Se and Bi were measured at energies of 100 and 200 keV, using scattering angles of 30°, 40°, 60° and 85°. The experimental cross sections were compared to matching theoretical Mott cross sections calculated by the same authors.

Lyman, E. M., Hanson, A. O. and Scott, M. B. (1951) *Phys. Rev.* **84**, 626.

Mott electron scattering measurements carried out on Al and Au at 15.7 MeV showed deviations due to finite nuclear size effects.

Maier, K. *et al.* (1988) *Z. Physik* **A330**, 173.

This experiment was designed as an "axion" search, but it shows the experimental ratio of Bhabha to Mott scattering at energies near 2300 keV.

McClelland, J. J. and Fink, M. (1985) *Phys. Rev.* **A31**, 1328.

Elastic and inelastic differential cross sections were measured over an angular range of 20 - 200 mrad (1° - 11°) for 35 keV electrons incident on Ne.

Missoni, G., Dick, C. E., Placious, R. C. and Motz, J. W. (1970) *Phys. Rev.* **A2**, 2309.

Measurements were made of inelastic electron scattering on C, Cu and Au at incident energies of 100, 200, 400 and 3000 keV.

Motz, J. W. and Placious, R. C. (1963a) *Phys. Rev.* **132**, 1120.

Measurements were made of the inelastic scattering of 500 keV electrons off Al and Au targets.

Motz, J. W., Placious, R. C. and Dick, C. E. (1963b) *Phys. Rev.* **132**, 2558.

Mott elastic scattering differential cross sections at electron energies of 50, 100, 200 and 400 keV, and over an angular range of 20° to 110°, were measured for thin targets of Au, Sn and Cu. Matching screening corrections were carried out by Lin (1964).

Neher, H. V. (1931) *Phys. Rev.* **38**, 1321.

Electron beams up to 145 keV in energy were used to investigate scattering from thin foils of Al, Ag and Au.

Niedrig, H. and Sieber, P. (1971) *Z. Angew. Phys.* **31**, 27.

The backscattering of electrons from various metals was investigated at energies between 20 and 53 keV. These measurements are relevant for electron microscopy.

Oms, J., Erman, P. and Hultberg, S. (1969) *Arkiv Fysik* **39**, 573.

A study was made of plural and multiple scattering for electron scattering from C, Al, Cu, Ag, and Pb at energies between 100 keV and 1 MeV . The experimental results were compared to various multiple scattering theories.

Paul W. and Reich, H. (1952) *Z. Physik* **131**, 326.

Relative cross sections were measured for electron scattering at 2.2 MeV on Al, Cu, Ag, Sn and Pt.

Pettus, W. G., Blosser, H. G. and Hereford, F. (1956) *Phys. Rev.* **101**, 17.

The Mott elastic scattering of 200 keV electrons on Au was measured at scattering angles between 70 and 150 degrees.

Randels, R. B., Chao, K. T. and Crane, H. R. (1945) *Phys. Rev.* **68**, 64.

Electrons with energies between 0.9 and 12 MeV were used in cloud chamber measurements to demonstrate that Ar, Kr, Xe, and air all give scattering cross sections which are in reasonable agreement with Mott scattering theory.

Rester, D. H. and Rainwater, W. J., Jr. (1965a) *Phys. Rev.* **138**, A12.

Elastic scattering measurements with 1 MeV electrons were made on thin aluminum foils at scattering angles from 20° to 150°.

Rester, D. H. and Rainwater, W. J., Jr. (1965b) *Phys. Rev.* **140**, A165.

Elastic differential cross sections at angles of 30° to 150° were measured on thin aluminum foils for electron scattering at energies of 100, 200, 500, 700, 1000, 1250, 1500, 2000, 2500 and 3000 keV.

Rester, D. H. and Edmonson, N. (1971) *Phys. Rev.* **A4**, 147.

The inelastic scattering of 200, 1000 and 2000 keV electrons from Al and Au targets was investigated.

Ruane, T. F., Waldman, B. and Miller, W. C. (1955) *Phys. Rev.* **98**, 1166A.

This A. P. S. abstract describes the scattering of 1.75 MeV electrons from thin foils of Al, Ni and Au.

Schmoranzer, H., Grabe, H. and Schiewe, B. (1975) *Appl. Phys. Lett.* **26**, 483.

The backscattering of multi-keV electrons from gold foils of various thicknesses was investigated experimentally, and its relationship to electron microscopy was reviewed.

Spiegel, V., Jr., Waldman, B. and Miller, W. C. (1955) *Phys. Rev.* **100**, 1244.

This A. P. S. abstract describes the scattering of 1.0, 1.75 and 2.5 MeV electrons from thin foils of Al, Ni, Ag and Au.

Spiegel, V., Jr., Ruane, T. F., Anthony, D. J., Waldman, B. and Miller, W. C. (1959) *Annals of Physics* **6**, 70.

Absolute Mott differential cross section measurements were made on Al, Ni, Ag and Au at electron energies of 1.0, 1.75 and 2.5 MeV. A review is given of previous Mott scattering experiments.

Tsertos, H. *et al.* (1988a) *Phys. Lett.* **B207**, 273.

This experiment was a search for resonant Bhabha scattering near 1.8 MeV (no resonance was found), and it gives the ratio of Bhabha to Mott scattering in this energy region.

Tsertos, H. *et al.* (1988b) *Z. Phys.* **A331**, 103.

This is the same experiment as Tsertos (1988a), and it has figures showing the Bhabha and Mott scattering events plotted separately.

Ulsh, R. C., Wellenstein, H. F. and Bonham, R. A. (1974) *J. Chem. Phys.* **60**, 103.

Electrons at 25 keV were used to study elastic and inelastic scattering effects on H_2.

Van de Graaff, R. J., Buechner, W. W. and Feshbach, H. (1946) *Phys. Rev.* **69**, 452.

Electrons from 1.27 to 2.27 MeV were scattered from Al, Cu, Ag, Pt, and Au foils of various thicknesses.

Wellenstein, H. F., Bonham, R. A. and Ulsh, R. C. (1973) *Phys. Rev.* **A8**, 304.

Elastic and inelastic differential scattering cross sections for 25 keV electrons on He were measured at scattering angles up to 10°.

Wong, T. C., Lee, J. S. and Bonham, R. A. (1975) *Phys. Rev.* **A11**, 1963.

Inelastic scattering measurements were made for 25 keV electrons incident on Ne, Ar, Kr and Xe.

Appendix B

Papers on Mott Double Scattering from Atomic Nuclei

In Appendix B, theoretical and experimental papers that relate to Mott double-scattering spin polarization and asymmetry experiments are summarized. Some of these papers also relate to Mott single scattering cross section experiments, and hence were included in Appendix A.

A. Theoretical Papers

Braicovich, L. and De Michelis, B. (1968a) *Nuovo Cimento* **58B**, 269.

The effect on Mott asymmetry experiments of multiple scattering in thick foils was investigated theoretically (see Braicovich 1968b for experimental results).

Bühring, W. (1968) *Z. Physik* **212**, 61.

Screened Mott elastic cross sections and asymmetry functions were calculated for Hg at 46, 79, 100 and 204 keV over the angular range 45° - 180°.

Farago, P. S. (1977) *Comments Atom. Mol. Phys.* **6**, 99.

Analogies and contrasts between optical polarizers and Mott spin polarizers are discussed.

Felsner, G. and Rose, M. E. (1961) *Nuovo Cimento* **20**, 1821.

The effect of inelastic scattering on the Mott asymmetry function for Au was calculated and found to be very small.

Fink, M. and Yates, A. C. (1970) *Atomic Data* **1**, 385. Part I.

The Dirac equation was used to calculate Mott elastic scattering and polarization for selected elements, using incident electron energies ranging from 100 to 1500 eV.

Fink, M. and Ingram, J. (1972) *Atomic Data* **4**, 129. Part II.

These are the same calculations as described in Fink (1970), but for different elements and wider (backscattering) angles.

Fletcher, G. D., Gay, T. J. and Lubell, M. S. (1968) *Phys. Rev.* **A34**, 911.

An analysis was made of Mott electron polarimeters, showing fundamental limits on the accuracy that can be achieved.

Gellrich, A., Jost, K. and Kessler, J. (1990) *Rev. Sci. Instrum.* **61**, 3399.

A discussion is given of instrumental asymmetries in Mott polarimeters.

Goeke, J., Kessler, J. and Hanne, G. F. (1987) *Phys. Rev. Lett.* **59**, 1413.

The use of circularly polarized He radiation for electron polarimetry is assessed.

Goerss, H.-J., Nordbeck, R.-P. and Bartschat, K. (1991) *J. Phys. B: At. Mol. Opt. Phys.* **24**, 2833.

Generalized STU parameters were calculated for the elastic and inelastic scattering of electrons from thallium atoms. Some comparisons to experimental data were made.

Gregory, D. and Fink, M. (1974) *Atomic Data and Nuclear Data Tables* **14**, 39. Part III.

Dirac equation calculations for electron scattering and spin polarization are given for an additional series of elements. See Fink (1970) and Fink (1972).

Hnizdo, V. (1973) *Nucl. Inst. Meth.* **109**, 503.

Monte Carlo calculations were carried out to delineate multiple scattering effects in Mott polarimeters.

Hodge, L. A., Moravec, T. J., Dunning, F. B. and Walters, G. K. (1979) *Rev. Sci Inst.* **55**, 5.

A new Mott polarimeter design is described.

Holzwarth, G. and Meister, H. J. (1964) *Nucl. Phys.* **59**, 56.

Screened Mott cross sections and asymmetry functions were calculated for Au and Hg over a range of energies from 0.2 to 290 keV.

Ivanovic, I. D. (1988) *J. Phys. A: Math. Gen.* **21**, 1857.

This is a mathematical discussion of Mott scattering spin measurements as a class of generalized quantum measurements.

Kelsey, E. J. (1979) *Phys. Rev.* **A19**, 1062.

The scattering of polarized electrons from H is discussed.

Kessler, J. (1976) *Polarized Electrons*, Springer-Verlag, Heidelberg.

This book gives a clear development of the Dirac equations as applied to Mott scattering and polarization experiments.

Khare, S. P., Kumar, A. and Vijayshri (1985a) *J. Phys. B: Atom. Mol. Phys.* **18**, 1827.

Optical model calculations of the spin polarization parameters S, T and U were carried out for electron scattering on Ca.

Lin, S., Sherman, N. and Percus, J. K. (1963) *Nucl. Phys.* **45**, 492.

Screened Mott elastic scattering cross sections and asymmetry functions were calculated for Au at 121 keV and for Hg at 46 and 79 keV, over a range of scattering angles from 15° to 165°.

Lin, S. (1964) *Phys. Rev.* **133**, A965.

Screened Mott elastic scattering cross sections and asymmetry functions were calculated for Au and Cu at energies from 50 to 400 keV, and for Hg at 204 keV, over a range of scattering angles from 10° to 170°.

McEachran, R. P. and Stauffer, A. D. (1986) *J. Phys. B: Atom. Mol. Phys.* **19**, 3523.

Calculations were made of the electron spin polarization for scattering from Xe at energies up to 300 eV.

Meister, H. J. and Weiss, H. F. (1968) *Z. Physik* **216**, 165.

The non-relativistic (Schrödinger) and relativistic (Dirac) equations are compared at low energies (below 200 eV) in order to assess the importance of relativistic effects on differential cross sections and spin polarizations.

Mohr, C. B. O. and Tassie, L. J. (1954) *Proc. Phys. Soc.* **A67**, 711.

Screened Mott elastic scattering cross sections and asymmetry functions were calculated for Au over an energy range from 0.9 to 121 keV.

Mott, N. F. and Massey, H. S. W. (1949) *The Theory of Atomic Collisions*, Second edition, Clarendon Press, Oxford.

Pages 74-87 of this reference give the basic Dirac equations for electron and positron Mott scattering and polarization.

Motz, J. W., Olson, H. and Koch, H. W. (1964) *Rev. Mod. Phys.* **36**, 881.

This is an extensive review of the calculations that have been made with respect to the Mott elastic scattering and polarization of electrons by atomic nuclei. Screening effects are discussed in detail. Numerical values for calculations by various authors are given.

Nahar, S. N. and Wadehra, J. M. (1991) *Phys. Rev.* **A43**, 1275.

Calculations of Mott scattering cross sections and P, T and U polarization vectors were carried out for electron and positron scattering off Ar at energies up to 300 eV, using a detailed treatment of the Dirac equation. Comparisons were made to experiment.

Rose, M. E. (1940) *Phys. Rev.* **57**, 280.

A review is given of the basic Mott equations for the scattering and polarization of electrons, and the experimental results up to that time (which were not very accurate) are discussed.

Schopper, H. (1959) *Nucl. Phys.* **11**, 675.

Calculations were made of Mott polarization effects in the plane of the scattering.

Sherman, N. (1956) *Phys. Rev.* **103**, 1601.

The Dirac equation was used to calculate unscreened Mott cross sections and asymmetry functions for Al, Cd and Hg at a variety of electron energies.

Sherman, N. and Nelson, D. F. (1959) *Phys. Rev.* **114**, 1541.

Unscreened Mott cross sections and asymmetry functions were calculated for Au and Al at 75 and 121 keV.

Tang, F.-C., Zhang, X., Dunning, F. B. and Walters, G. K. (1988) *Rev. Sci. Inst.* **59**, 504.

A compact low-energy Mott polarimeter is described.

Tolhoek, H. A. (1956) *Rev. Mod. Phys.* **28**, 277.

A detailed description is given of Mott polarization experiments from both the theoretical and experimental viewpoints.

Walker, D. W. (1971) *Advances in Physics* **20**, 257.

The use of the Dirac equation in calculating Mott low energy electron elastic scattering and polarization is discussed in detail.

B. Experimental Papers Below 10 keV

Bartschat, K., Hanne, G. F., Wolcke, A. and Kessler, J. (1981) *Phys. Rev. Lett.* **47**, 997.

Spin polarization and asymmetry measurements were made for electron scattering from Hg atoms at energies up to 8 eV.

Bartschat, K., Blum, K., Burke, P. G., Hanne, G. F. and Scott, N. S. (1984) *J. Phys. B: At. Mol. Phys.* **17**, 3797.

Calculations were made of the spin polarization and asymmetry for the scattering of 1.9 and 3.5 eV electrons from Hg and Ta, and comparisons were made to experiment.

Bartschat, K., Goerss, H.-J. and Nordbeck, R.-P. (1990) *Z. Physik* **D17**, 25.

The S, T, U asymmetry parameters for Th and Pb were calculated at energies up to 5 eV and compared to experiment.

Berger, O. and Kessler, J. (1986) *J. Phys. B: At. Mol. Phys.* **19**, 3539.

The angular distribution of the asymmetry parameters S, T, and U was measured for Hg and Xe, using energies up to 350 eV for Xe, and up to 150 eV for Hg.

Deichsel, H. and Reichert, E. (1964) *Phys. Lett.* **13**, 125.

Mott asymmetry measurements on a beam of Hg atoms were carried out at 900 eV.

Deichsel, H. and Reichert, E. (1965) *Z. Physik* **185**, 169.

Mott asymmetry measurements on a beam of Hg atoms were carried out at 0.9, 1.2 and 1.5 keV.

Eckstein, W. (1967) *Z. Physik* **203**, 59.

The polarization of electrons elastically scattered from a solid Hg target was studied for electron energies between 300 and 900 eV.

Geesmann, H., Bartsch, M., Hanne, G. F. and Kessler, J. (1991) *J. Phys. B: At. Mol. Opt. Phys.* **24**, 2817.

The Mott asymmetry for polarized electrons scattering from Th and Pb atoms was measured at energies between 2.5 and 14 eV.

Jost, K. and Kessler, J. (1966) *Z. Physik* **195**, 1.

Polarization measurements were made of electron scattering from a mercury-atom beam for incident electron energies from 0.18 to 1.7 keV.

Kelley, M. H. (1989) *AIP Conference Proceedings* **205**, 616.

A discussion is given of spin-polarization measurements in electron-atom scattering.

Kessler, J. (1969) *Rev. Mod. Phys.* **41**, 3.

A general review is given of electron spin polarization experiments up to a few keV.

Loth, R. (1967) *Z. Physik* **203**, 66.

The polarization of 900 eV electrons elastically scattered from solid W, Pt, and Au targets was investigated.

Mehr, J. (1967) *Z. Physik* **198**, 345.

Elastic scattering angular distributions were measured for 5 to 1000 eV electrons incident on He, Ne, Ar, Kr, and Xe, and the spin polarization angular distribution was measured for 40 eV electrons incident on Ar.

Schackert, K. (1968) *Z. Physik* **213**, 316.

The angular dependence of the spin polarization produced by 40 to 150 eV electrons scattering on He, Ne, Ar, Kr, and Xe was measured.

Wainwright, P. F., Alguard, M. J., Baum, G. and Lubell, M. S. (1978) *Rev. Sci. Instrum.* **49**, 571.

A Fano source was used to produce polarized electrons for use in low-energy atomic spin measurements.

C. Experimental Papers Above 10 keV

Apalin, V. A. *et al.* (1962) *Nucl. Phys.* **31**, 657.

Mott double scattering at angles of 120° on Au was used to measure the electron asymmetry function over an energy range from 45 to 245 keV.

Bienlein, H., Felsner, G., Günther, K., von Issendorff, H. and Wegener, H. (1959a) *Z. Physik* **154**, 376.

The Mott asymmetry at scattering angles of 120° was measured for Au, Ce, Ag and Cu at 120, 155 and 209 keV.

Bienlein, H. *et al.* (1959b) *Z. Physik* **155**, 101.

The angular dependence of the asymmetry of Mott scattering for Au at 155 keV was measured for scattering angles from 40° to 150°.

Bienlein, H., Günther, K., von Issendorff, H. and Wegener, H. (1959c) *Nucl. Inst. Meth.* **4**, 79.

An experimental arrangement is described for measuring longitudinal electron polarization by Mott scattering.

Boersch, H., Schliepe, R. and Schriefl, K. E. (1971) *Nucl. Phys.* **A163**, 625.

Mott 120° double scattering measurements from 22 to 48 keV were carried out using thick gold targets.

Braicovich, L. and De Michelis, B. (1968b) *Nuovo Cimento* **58B**, 460.

The multiple-scattering theory of Braicovich (1968a) was tested by carrying out electron double scattering measurements on thick Au targets at energies ranging from 255 to 1380 keV.

Brosi, A. R., Galonsky, A. I., Ketelle, B. H. and Willard, H. B. (1962) *Nucl. Phys.* **33**, 353.

Mott scattering was used to determine the longitudinal polarization of 616 keV electrons from the beta decay of P^{32}.

Campbell, D. M., Herman, C., Lampel, G. and Owen, R. (1985) *J. Phys. E* **18**, 664.
Measurements of the Sherman function S were carried out for gold at energies ranging from 20 to 100 keV.

Eckardt, V., Ladage, A. and Moellendorff, U. V. (1964) *Phys. Lett.* **13**, 53.
Mott asymmetry in electron double-scattering from gold was measured at 100 keV, and the longitudinal polarization of electrons from the beta decay of Co^{60} and Pm^{147} was measured from 40 to 210 keV.

Gauder, R. *et al.* (1988) *Z. Physik* **A330**, 423.
Nuclear beta decays were measured to see if the longitudinal polarization accurately follows the expected v/c dependence. No deviations were found for 80 - 200 keV electrons.

Gray, L. G., Hart, M. W., Dunning, F. B. and Walters, G. K. (1984) *Rev. Sci. Inst.* **55**, 88.
The Sherman function S(120°) for a gold target was measured from 20 to 120 keV.

Greenberg, J. S., Malone, D. P., Gluckstern, R. L. and Hughes, V. W. (1960) *Phys. Rev.* 120, 1393.
Mott scattering was used to analyze the longitudinal polarization of beta rays emitted from various radioactive nuclei.

Hatamian, M. S., Coulter, K. and Skalsey, M. (1985) *Phys. Rev.* **C31**, 2222.
A Mott polarimeter was used to look for time-reversal-violating polarization components in beta decay. None were found.

Krause, H. H., Arnold, W., Berg, H., Ulbricht, J. and Clausnitzer, G. (1979) *Z. Physik* **A293**, 343.
The Mott-Schwinger interaction, which gives nucleon polarizations via Mott-type magnetic scattering, was verified for the case of 450 to 600 keV polarized protons scattered from ^{12}C.

Kulkarni, V. G., Panat, C. V. and Thosar, B. V. (1963) *Phys. Lett.* **7**, 52.
The longitudinal polarization of the electrons emitted in the beta decay of Cd^{115m} was measured.

Lazarus, D. M. and Greenberg, J. S. (1970) *Phys. Rev.* **D2**, 45.
The v/c dependence of the electron polarization from the beta decay of Co^{60} was checked and found to be accurate.

McClelland, J. J., Scheinfein, M. R. and Pierce, D. T. (1989) *Rev. Sci. Instrum.* 60, 683.
The Sherman function S(120°) was measured for thorium and gold over the energy range from 20 to 100 keV.

Mikaelyan, L., Borovoi, A. and Denisov, E. (1963) *Nucl. Phys.* **47**, 328.

The Mott asymmetry function for 45 to 245 keV electron double scattering on Au was measured at scattering angles from 90° to 150°.

Nelson, D. F. and Pidd, R. W. (1959) *Phys. Rev.* **114**, 728.

The Mott asymmetry function for 121 keV electron double scattering on Au was measured over a range of scattering angles.

Pettus, W. G. (1958) *Phys. Rev.* **109**, 1458.

The polarization of electrons double-scattered by thin gold foils was measured for energies ranging from 80 to 190 keV, and scattering angles of 60°, 90° and 120°.

Ryu, N. (1952a) *J. Phys. Soc. Japan* **7**,125.

Mott asymmetry measurements were carried out on Au for electron energies of 60 to 130 keV, and using equal-angle scatterings.

Ryu, N. (1952b) *J. Phys. Soc. Japan* **7**, 130.

The Mott asymmetry measurements of Ryu (1952a) were extended to include unequal angles in the two scatterings.

Ryu, N. (1953) *J. Phys. Soc. Japan* **8**, 804.

The measurements described in Ryu (1952a,b) were improved and extended.

Shull, C. G., Chase, C. T. and Myers, F. E. (1943) *Phys. Rev.* **63**, 29.

This experiment on Au at 400 keV gave the first clear observation of Mott asymmetry in electron double scattering.

Spivak, P. E., Mikaelyan, L. A., Kutikov, I. Ye. and Apalin, V. F. (1961) *Nucl. Phys.* 23, 169.

Mott asymmetry values were measured for gold at 245 and 290 keV, and Mott scattering was used to determine longitudinal polarizations for the electrons from various beta-decay sources.

Van Duiden, R. J. and Aalders, J. W. G. (1968) *Nucl. Phys.* **A115**, 353.

The triple scattering of 261 keV electrons on Au was measured.

Van Klinken, J. (1964) *Nucl. Phys.* **58**, 283.

Mott scattering was used to measure longitudinal beta ray polarizations.

Van Klinken, J. (1966a) *Nucl. Phys.* **75**, 145.

Mott scattering was used to measure longitudinal beta ray polarizations from 46 to 261 keV.

Van Klinken, J. (1966b) *Nucl. Phys.* 75, 161.
The Sherman function S was measured for Au at energies from 26 to 261 keV.

Wegener, H. (1958) *Z. Physik* **151**, 252.
Mott polarization measurements on Au were studied.

Bibliography

Aharoni, J. (1965) *The Special Theory of Relativity*, Clarendon, Oxford.

Amendolia, S. R. *et al.* (1984) *Phys. Lett.* **146B**, 116.

Amendolia, S. R. *et al.* (1986a) *Phys. Lett.* **178B**, 435.

Amendolia, S. R. *et al.* (1986b) *Nucl. Phys.* **B277**, 168.

Anderson, C. D. (1933) *Phys. Rev.* **43**, 481.

Anthony, D. J., Waldman, B. and Miller, W. C. (1953) *Phys. Rev.* **91**, 439a.

Apalin, V. F. *et al.* (1962) *Nucl. Phys.* **31**, 657.

Arzelies, H. (1966) *Relativistic Kinematics*, Pergamon, Oxford.

Ashkin, A. *et al.* (1954) *Phys. Rev.* **94**, 357.

Atwood, D. K., Horne, M. A., Shull, C. G. and Arthur, J. (1984) *Phys. Rev. Lett.* **52**, 1673.

Bailey, J. *et al.* (1979) *Nucl. Phys.* **B150**, 1.

Barber, A. and Champion, F. C. (1938) *Proc. Roy. Soc.* **A168**, 159.

Barber W. C. *et al.* (1953) *Phys. Rev.* **89**, 950.

Bargmann, V., Michel, L. and Telegdi, V. L. (1959) *Phys. Rev. Lett.* **2**, 435.

Bartschat, K., Hanne, G. F., Wolcke, A. and Kessler, J. (1981) *Phys. Rev. Lett.* **47**, 997.

Barut, A. O. (ed) (1980) *Foundations of Radiation Theory and Quantum Electrodynamics*, Plenum, New York.

Barut, A. O. and Bracken, A. J. (1981) *Phys. Rev.* **D24**, 3333.

Barut, A. O. and Zanghi, N. (1984) *Phys. Rev. Lett.* **52**, 2009.

Barut, A. O. and Unal, N. (1989a) *Phys. Rev.* **A40**, 5404.

Barut, A. O. and Dowling, J. P. (1989b) *Z. Naturforsch.* **44a**, 1051.

Barut, A. O. (1991) in D. Hestenes and A. Weingartshofer (eds).

Bayard, R. T. and Yntema, J. L. (1955) *Phys. Rev.* **97**, 372.

Behrends, F. A. *et al.* (1974) *Nucl. Phys.* **B68**, 541.

Behrends, F. A. and Kleiss, R. (1981) *Nucl. Phys.* **B177**, 237.

Behrends, F. A. *et al.* (1982) *Nucl. Phys.* **B202**, 63.

Belinfante, F. J. (1939) *Physica* **6**, 887.

Belloni, L. (1981) *Lett. Nuovo Cimento* **31**, 131.

Bender, D. *et al.* (1984) *Phys. Rev.* **D30**, 515.

Berger, O. and Kessler, J. (1986) *J. Phys. B: At. Mol. Phys.* **19**, 3539.

Bergmann, P. G. (1942) *Introduction to the Theory of Relativity*, Prentice-Hall, Englewood Cliffs.

Bethe, H. A. (1947) *Phys. Rev.* **72**, 339.

Bhabha, H. J. (1936) *Proc. Royal Soc. (London)* **A154**, 195.

Bialynicki-Birula, I. (1982) *Phys. Rev.* **D28**, 2114.

Bienlein, H. *et al.* (1959a) *Z. Physik* **154**, 376.

Bienlein, H. *et al.* (1959b) *Z. Physik* **155**, 101.

Blanco, R., Pesquera, L. and Jimenez, J. L. (1986) *Phys. Rev.* **D34**, 452.

Blanco, R. (1987) *J. Phys. A: Math. Gen.* **20**, 5885.

Blatt, J. M. and Weisskopf, V. F. (1952) *Theoretical Nuclear Physics*, Wiley, New York.

Blokhintsev, D. I. (1973) *Space and Time in the Microworld*, Reidel, Dordrecht.

Boersch, H., Schliepe, R. and Schriefl, K. E. (1971) *Nucl. Phys.* **A163**, 625.

Bohm, D. and Weinstein, M. (1948) *Phys. Rev.* **74**, 1789.

Bohm, D. (1965) *The Special Theory of Relativity*, Benjamin, New York.

Born, M. and Schrödinger, E. (1935) *Nature (London)* **135**, 342.

Born, M. (1962) *Einstein's Theory of Relativity*, Dover, New York.

Boyer, R. H. (1962) *Nature* **196**, 886.

Boyer, T. H. (1968) *Phys. Rev.* **174**, 1764.

Boyer, T. H. (1969) *J. Math. Phys.* **10**, 1729.

Boyer, T. H. (1980) in A. O. Barut (ed).

Boyer, T. H. (1982) *Phys. Rev.* **D25**, 3246.

Boyer, T. H. (1984) *Phys. Rev.* **D29**, 1089.

Boyer, T. H. (1985a) *Am. J. Phys.* **53** (2), 167.

Boyer, T. H. (1985b) *Scientific American* **251**, 8, 70.

Braicovich, L., De Michelis, B. and Fasana, A. (1967) *Phys. Rev.* **154**, 234.

Braicovich, L. and De Michelis, B. (1968) *Nuovo Cimento* **58B**, 460.

Bransden, B. H. and McDowell, R. C. (1978) *Physics Reports* **46**, 249.

Breit, G. (1947) *Phys. Rev.* **72**, 984.

Bridgman, P. W. (1962) *A Sophisticate's Primer of Relativity*, Wesleyan, Middletown.

Bromberg, J. P. (1969) *J. Chem. Phys.* **51**, 4117.

Bromberg, J. P. (1974) *J. Chem. Phys.* **61**, 963.

Bronshtein, I. M. and Pronin, V. P. (1976a) *Sov. Phys. Solid State* **17**, 1363.

Bronshtein, I. M. and Pronin, V. P. (1976b) *Sov. Phys. Solid State* **17**, 1610.

Bronshtein, I. M. and Pronin, V. P. (1976c) *Sov. Phys. Solid State* **17**, 1672.

Brown, B., Matsukawa, E. and Stewardson, E. A. (1956) *Proc. Phys. Soc.* **69**, 496.

Buckman, S. J., Noble, C. J. and Teubner, P. J. O. (1979) *J. Phys. B: Atom. Mol. Phys.* **12**, 3077.

Buechner, W. W. *et al.* (1947) *Phys. Rev.* **72**, 678.

Bühring, W. (1968) *Z. Physik* **212**, 61.

Bunge, M. (1955) *Nuovo Cimento* **1**, 977.

Burge, R. E. and Smith, G. H. (1962) *Proc. Phys. Soc.* **79**, 673.

Cagnac, B. and Pebay-Peroula, J. C. (1971) *Modern Atomic Physics: Fundamental Principles*, Wiley, New York.

Caldirola, P. (1956) *Suppl. Nuovo Cimento* **3**, 297.

Caldirola, P., Casati, G. and Prosperetti, A. (1978) *Nuovo Cimento* **43A**, 127.

Calmet, J., Narison, S., Perrottet, M. and de Rafael, E. (1977) *Rev. Mod. Phys.* **49**, 21.

Cameron, A. G. W. (1971) *The Crab Nebula*, R. D. Davies and F. G. Smith (eds), Reidel, Dordrecht.

Campbell, D. M. *et al.* (1984) *J. Phys. E: Sci. Instrum.* **18**, 664.

Carmeli, M. (1984a) *Lett. Nuovo Cimento* **41**, 545.

Carmeli, M. (1984b) *Lett. Nuovo Cimento* **41**, 551.

Carmeli, M. (1985) *Lett. Nuovo Cimento* **42**, 67.

Casimir, H. B. G. (1948) *Proc. Kon. Ned. Akad. Wetenschap* **51**, 793.

Champion, F. C. and Barber, A. (1939) *Phys. Rev.* **55**, 111.

Chapman, K. R. *et al.* (1955) *Proc. Phys. Soc.* **68**, 928.

Chase, C. T. and Cox, R. T. (1940) *Phys. Rev.* **58**, 243.

Coleman, J. A. (1958) *Relativity for the Layman,* Macmillan, New York.

Coleman, P. G. and McNutt, J. D. (1979) *Phys. Rev. Lett.* **42**, 1130.

Corben, H. C. (1968) *Classical and Quantum Theories of Spinning Particles*, Holden-Day, San Francisco.

Cosslett, V. E. and Thomas, R. N. (1964a) *Brit. J. Appl. Phys.* **15**, 883.

Cosslett, V. E. and Thomas, R. N. (1964b) *Brit. J. Appl. Phys.* **15**, 1283.

Cosslett, V. E. and Thomas, R. N. (1965) *Brit. J. Appl. Phys.* **16**, 779.

Costa de Beauregard, O. (1966) *Precis of Special Relativity*, Academic Press, New York.

Cox, R. T. and Chase, C. T. (1937) *Phys. Rev.* **51**, 141.

Crawford, J. F. *et al.* (1988) *Phys. Lett.* **B213**, 391.

Cullwick, E. G. (1958) *Electromagnetism and Relativity*, Longmans, Green, London.

Cushing, J. T. (1981) *Am. J. Phys.* **49**, 1133.

Cvijanovich, G. B. and Vigier, J.-P. (1977) *Found. Phys.* **7**, 77.

Daboul, J. and Jensen, J. H. D. (1973) *Z. Physik*, **265**, 455.

Daboul, J. (1975) *Z. Physik* **B21**, 115.

Damodaran, K. K. and Curr, R. M. (1956) *Proc. Phys. Soc.* **69**, 196.

Danjo, A. (1988) *J. Phys. B: At. Mol. Opt. Phys.* **21**, 3759.

Davies, P. C. W. (1975) *J. Phys.* (GB) **A8**, 609.

De Broglie, L. (1924) *Doctoral Thesis*, published in *Ann. Phys. (Paris) (10)* **III**, 22 (1925).

De Broglie, L. (1976) *Ann. Fond. Louis de Broglie* **1**, 116.

Deichsel, H. and Reichert, E. (1964) *Phys. Lett.* **13**, 125.

Deichsel, H. and Reichert, E. (1965) *Z. Physik* **185**, 169.

De la Peña, L., Jimenez, J. L. and Montemayor, R. (1982) *Nuovo Cimento* **69B**, 71.

De la Peña, L. (1983) in *Stochastic Processes Applied to Physics and other Related Fields*, B. Gomez, S. M. Moore, A. M. Rodriguez-Vargas, and A. Rueda (eds), World Scientific, Singapore.

Delfino, M. C. (1985) thesis, University of Wisconsin - Madison.

Dewan, E. and Beran, M. (1959) *Am. J. Phys.* **27**, 517.

Dewan, E. M. (1963) *Am. J. Phys.* **31**, 383.

Dingle, H. (1940) *The Special Theory of Relativity*, Metheun, London.

Dirac, P. A. M. (1928) *Proc. Roy. Soc. (London)* **A117**, 610.

Doggett, J. A. and Spencer, L. V. (1956) *Phys. Rev.* **103**, 1597.

Drell, S. D. (1958) *Annals of Physics* **4**, 75.

DuBois, R. D. and Rudd, M. E. (1976) *J. Phys. B: Atom. Mol. Phys.* **9**, 2657.

Duguet, A., Bennani, A. and Roualt, M. (1983) *J. Chem. Phys.* **79**, 2786.

Duval, Ch. (1975) CNRS, Marseille report 75/P.767.

Dyson, F. J. (1951) *"Advanced Quantum Mechanics"*, lecture notes, Cornell University, unpublished, 2nd edition, edited by M. J. Moravcsik.

Eckardt, V., Ladage, A. and Moellendorff, U. V. (1964) *Phys. Lett.* **13**, 53.

Eckstein, W. (1967) *Z. Physik* **203**, 59.

Eichten, E. J. *et al.* (1983) *Phys. Rev. Lett.* **50**, 811.

Einstein, A. *et al.* (1923) *The Principle of Relativity*, Dover, New York.

Erber, T. (1961) *Fortschritte der Physik* **9**, 343.

Erman, P. (1970) *Physica Scripta* **1**, 93.

Essen, L. (1971) *The Special Theory of Relativity: A Critical Analysis*, Clarendon, Oxford.

Feld, B. T. (1969) *Models of Elementary Particles*, Blaisdell, Waltham.

Fermi, E. (1932) *Rev. Mod. Phys.* **4**, 87.

Feynman, R. P. (1961a) in *The Quantum Theory of Fields, Proceedings of the Twelfth Conference on Physics at the University of Brussels, October, 1961*, R. Stoops (ed), Interscience, New York.

Feynman, R. P. (1961b) *Theory of Fundamental Processes*, Benjamin, New York.

Feynman, R. P., Leighton, R. B. and Sands, M. (1964) *The Feynman Lectures in Physics*, Addison-Wesley, Reading.

Feynman, R. P. (1985) *QED, The Strange Theory of Light and Matter*, Princeton University Press, Princeton.

Fink, M. and Kessler, J. (1966) *Z. Physik* **196**, 1.

Fink, M. and Bonham, R. A. (1969) *Phys. Rev.* **187**, 114.

Fink, M. and Moore, P. G. (1977) *Phys. Rev.* **A15**, 112.

Fletcher, G. D., Gay, T. J. and Lubell, M. S. (1968) *Phys. Rev.* **A34**, 911.

Fletcher, J. G. (1963) *Nature* **199**, 994.

Fock, V. (1964) *The Theory of Space, Time, and Gravitation*, Macmillan, New York.

Fokker, A. D. (1929) *Z. Physik* **58**, 386.

Franca, H. M., Marques, G. C. and da Silva, A. J. (1978) *Nuovo Cimento* **48A**, 65.

French, A. P. (1958) *Principles of Modern Physics*, Wiley, New York.

French, A. P. (1968) *Special Relativity*, Norton, New York.

Fryberger, D. (1975) *SLAC-PUB-1611*, July.

Gardner, G. H. F. (1952) *Nature* **170**, 243.

Geesmann, H., Bartsch, M., Hanne, G. F. and Kessler, J. (1991) *J. Phys. B: At. Mol. Opt. Phys.* **24**, 2817.

Gergely, G. (1987) *Vacuum* **37**, 149.

Goldberg, S. (1984) *Understanding Relativity,* Birkhäuser, Boston.
Goldstein, H. (1950) *Classical Mechanics,* Addison-Wesley, Cambridge.
Gottfried, K. and Weisskopf, V. F. (1984) *Concepts of Particle Physics, Volume 1,* Clarendon Press, Oxford.
Gottfried, K. and Weisskopf, V. F. (1986) *Concepts of Particle Physics, Volume II,* Clarendon Press, Oxford.
Grachev, B. D. *et al.* (1978) *Sov. Tech. Phys. Lett.* **4** (3), 110.
Grachev, B. D. *et al.* (1980) *Sov. Phys. JETP* **52** (5), 827.
Grandy, W. T., Jr. and Aghazadeh, A. (1982) *Annals of Physics* **142**, 284.
Gray, L. G. *et al.* (1984) *Rev. Sci. Instrum.* **55**, 88.
Greenberg, J. S. *et al.* (1960) *Phys. Rev.* **120**, 1393.
Greenberger, D. M. (1983) *Rev. Mod. Phys.* **55**, 875.
Grøn, Ø. (1979) *Found. Phys.* **9**, 353.
Guggenheim, E. A. (1967) *Elements and Formulae of Special Relativity,* Pergamon, Oxford.
Gupta, G. P. and Mathur, K. C. (1979) *J. Phys. B: Atom. Mol. Phys.* **12**, 3071.
Gupta, S. C. and Rees, J. A. (1975) *J. Phys. B: Atom. Mol. Phys.* **8**, 1267.
Head, C. E. and Moore-Head, M. E. (1981) *Phys. Lett.* **106B**, 111.
Heitler, W. (1954) *The Quantum Theory of Radiation*, Third Edition, Oxford University Press, London.
Hestenes, D. and Weingartshofer, A. (eds) (1991) *The Electron: New Theory and Experiment,* Kluwer Academic Publishers, Dordrecht.
Hill, E. L. (1946) *Phys. Rev.* **69**, 488.
Hill, E. L. (1947) *Phys. Rev.* **71**, 318.
Hodge, L. A. *et al.* (1979) *Rev. Sci. Instrum.* **50**, 5.
Hoffmann, B. (1983) *Relativity and its Roots*, Freeman, New York.
Hogarth, J. E. and McCrea, W. H. (1952) *Proc. Camb. Phil. Soc.* **48**, 616.
Holtkamp, G., Jost, K., Peitzmann, J. and Kessler, J. (1987) *J. Phys. B: Atom. Mol. Phys.* **20**, 4543.
Holzwarth, G. and Meister, H. J. (1964) *Nucl. Phys.* **59**, 56.
Hurley, W. J. (1984) *Physics Today* **37**, 8, 80.
Hyder, G. M. A. *et al.* (1986) *Phys. Rev. Lett.* **57**, 2252.
Ichimura, S., Aratama, M. and Shimizu, R. (1980) *J. Appl. Phys.* **51**, 2853.
Iga, I. *et al.* (1987) *J. Phys. B: Atom. Mol. Phys.* **20**, 1095.
Jackson, J. D. (1962) *Classical Electrodynamics*, Wiley, New York.
Jammer, M. (1966) *The Conceptual Development of Quantum Mechanics*, Mc Graw-Hill, New York.
Janossy, L. (1971) *Theory of Relativity based on Physical Reality*, Akademiai Kiado, Budapest.
Jansen, R. H. J. *et al.* (1976a) *J. Phys. B: Atom. Mol. Phys.* **9**, 185.
Jansen, R. H. J. and de Heer, F. J. (1976b) *J. Phys. B: Atom. Mol. Phys.* **9**, 213.
Jauch, J. M. and Rohrlich, F. (1976) *The Theory of Photons and Electrons*, Second Edition, Springer-Verlag, New York.

Jaynes, E. T. (1991) in D. Hestenes and A. Weingartshofer (eds).

Jost, K. and Kessler, J. (1966) *Z. Physik* **195**, 1.

Kacser, C. (1967) *Introduction to the Special Theory of Relativity*, Prentice-Hall, Englewood Cliffs.

Kanter, H. (1964) *Brit. J. Appl. Phys.* **15**, 555.

Kaufmann, W. (1906) *Ann. Phys.* **19**, 487 and **20**, 639.

Kaup, D. J. (1966) *Phys. Rev.* **152**, 1130.

Kauppila, W. E. *et al.* (n.d.) in *Annihilation in Gases and Galaxies*, R. Drachman (ed), Greenbelt, MD: Goddard Space Research Center, p. 113.

Kauppila, W. E. and Stein, T. S. (1987) in *Atomic Physics with Positrons*, J. W. Humberston and E. A. G. Armour (eds), Plenum Press, New York, p. 27.

Kauppila, W. E. and Stein, T. S. (1989) in *AIP Conference Proceedings* **205**, p. 627.

Keck, K. (1962) *Z. Physik* **167**, 468.

Kelley, M. H. (1969) in *AIP Conference Proceedings* **205**, p. 103.

Kessler, J. (1959) *Z. Physik* **155**, 350.

Kessler, J. and Lindner, H. (1965) *Z. Physik* **183**, 1.

Kessler, J. and Weichert, N. (1968) *Z. Physik* **212**, 48.

Kessler, J. (1969) *Rev. Mod. Phys.* **41**, 3.

Kessler, J. (1976) *Polarized Electrons,* Springer-Verlag, Berlin.

Ketkar, S. N., Fink, M. and Bonham, R. A. (1983) *Phys. Rev.* **A27**, 806.

Kilmister, C. W. (1965) *The Environment in Modern Physics*, Elsevier, New York.

Kinoshita, T. and Lindquist, W. B. (1981) *Phys. Rev. Lett.* **47**, 1573.

Kinoshita, T., Nizic, B. and Okamoto, Y. (1984) *Phys. Rev. Lett.* **52**, 717.

Kinzinger, E. and Bothe, W. (1952) *Z. Naturforsch.* **7A**, 390.

Kinzinger, E. (1953) *Z. Naturforsch.* **8A**, 312.

Klarmann, H. and Bothe, W. (1936) *Z. Physik* **101**, 489.

Klein, A. G. and Werner, S. A. (1983) *Reports on Progress in Physics* **46**, 259.

Klewer, M., Beerlage, M. J. M. and van der Wiel, M. J. (1980) *J. Phys. B: Atom. Mol. Phys.* **13**, 571.

Kobayashi, T. and Shimizu, S. (1972) *J. Phys. B: Atom. Mol. Phys.* **5**, L211.

Kramers, H. A. (1958) *Quantum Mechanics*, North-Holland, Amsterdam.

Kroll, N. M. (1966) *Nuovo Cimento* **45A**, 65.

Kursunoglu, B. (1951) *Proc. Camb. Phil. Soc.* **47**, 177.

Lamb, W. E., Jr. and Retherford, R. C. (1947) *Phys. Rev.* **72**, 241.

Lamb, W. E., Jr. (1951) *Reports on Progress in Physics* **14**, 20.

Lautrup, B. E., Peterman, A. and de Rafael, E. (1972) *Physics Reports* **3C**, 193.

Lesiak, B., Jablonski, A. and Gergely, G. (1990) *Vacuum* **40**, 67.

Levine, H., Moniz, E. J. and Sharp, D. H. (1977) *Am. J. Phys.* **45** (1), 75.

Lichtenberg, D. B. (1962) *Nature* **196**, 886.

Lin, S., Sherman, N. and Percus, J. (1963) *Nucl. Phys.* **45**, 492.

Lin, S. (1964) *Phys. Rev.* **133**, A965.

Lopez, C. A. (1984) *Phys. Rev.* **D30**, 313.

Lorentz, H. A. (1952) *The Theory of Electrons*, Second Edition, Dover, New York.

Loth, R. (1967) *Z. Physik* **203**, 66.

Lucas, J. R. (1973) *A Treatise on Time and Space*, Metheun, London.

Lyman, E. M., Hanson, A. O. and Scott, M. B. (1951) *Phys. Rev.* **84**, 626.

Mac Gregor, M. H. and Wiedenbeck, M. L. (1952) *Phys. Rev.* **86**, 240.

Mac Gregor, M. H. and Wiedenbeck, M. L. (1954) *Phys. Rev.* **94**, 138.

Mac Gregor, M. H. (1970) *Lett. Nuovo Cimento* **4**, 211.

Mac Gregor, M. H. (1974) *Phys. Rev.* **D9**, 1259.

Mac Gregor, M. H. (1978) *The Nature of the Elementary Particle*, Springer-Verlag, Heidelberg.

Mac Gregor, M. H. (1981) *Lett. Nuovo Cimento* **30**, 417.

Mac Gregor, M. H. (1985a) *Lett. Nuovo Cimento* **43**, 49.

Mac Gregor, M. H. (1985b) *Lett. Nuovo Cimento* **44**, 697.

Mac Gregor, M. H. (1987) *Bull. A. P. S.* **32**, 1022.

Mac Gregor, M. H. (1988) *Found. Phys. Lett.* **1**, 25.

Mac Gregor, M. H. (1989) *Found. Phys. Lett.* **2**, 577.

Mac Gregor, M. H. (1990) *Nuovo Cimento* **103A**, 983.

Mac Gregor, M. H. (1992) *Found. Phys. Lett.* **5**, 15.

Mannheim, P. D. (1978) *Nucl. Phys.* **B143**, 285.

Margenau, H. (1961) in *Quantum Theory, I. Elements,* D. R. Bates (ed), Academic Press, New York.

Marinkovic, B. *et al.* (1991) *J. Phys. B: At. Mol. Opt. Phys.* **24**, 1817.

McClelland, J. J. and Fink, M. (1985) *Phys. Rev.* **A31**, 1328.

McClelland, J. J., Scheinfeld, M. R. and Pierce, D. T. (1989) *Rev. Sci. Instrum.* **60**, 683.

McFarlane, K. *et al.* (1980) *Int. J. Theor. Phys.* **19**, 347.

Mehr, J. (1967) *Z. Physik* **198**, 345.

Mercier, A. and Kervaire, M. (1956) (eds) *Jubilee of Relativity Theory, Helvetia Physica Acta, Supplement IV,* Birkhauser, Basel.

Mermin, N. D. (1968) *Space and Time in Special Relativity*, McGraw-Hill, New York.

Mikaelyan, L., Borovoi, A. and Denisov, E. (1963) *Nucl. Phys.* **47**, 328.

Miller, A. I. (1976) *Am. J. Phys.* **44**, 912.

Miller, A. I. (1981) *Albert Einstein's Special Theory of Relativity,* Addison-Wesley, Reading.

Milonni, P. W. (1980) in A. O. Barut (ed).

Mitroy, J., McCarthy, I. E. and Stelbovics, A. T. (1987) *J. Phys. B: Atom. Mol. Phys.* **20**, 4827.

Møller, C. (1932) *Annalen der Physik* **14**, 531.

Møller, C. (1952) *The Theory of Relativity*, Clarendon Press, Oxford.

Moniz, E. J. and Sharp, D. H. (1977) *Phys. Rev.* **D15**, 2850.

Moore, P. and Fink, M. (1972) *Phys. Rev.* **A5**, 1747.

Mott, N. F. and Massey, H. S. W. (1949) *The Theory of Atomic Collisions,* Second Edition, Oxford University Press, London.

Motz, J. W., Placious, R. C. and Dick, C. E. (1963) *Phys. Rev.* **132**, 2558.

Moyer, D. F. (1981) *Am. J. Phys.* **49**, 944.
Nafe, J. E., Nelson, E. B. and Rabi, I. I. (1947) *Phys. Rev.* **71**, 914.
Naon, M. and Cornille, M. (1972) *J. Phys. B: Atom. Mol. Phys.* **5**, 1965.
Naon, M., Cornille, M. and Kim, Y. (1975) *J. Phys. B: Atom. Mol. Phys.* **8**, 864.
Neher, H. V. (1931) *Phys. Rev.* **38**, 1321.
Nelson, D. F. and Pidd, R. W. (1959) *Phys. Rev.* **114**, 728.
Nernst, W. (1916) *Verh. Dtsch. Phys. Ges.* **18**, 83.
Neumann, G. (1914) *Ann. Phys.* **45**, 529.
Niedrig, H. and Sieber, P. (1971) *Z. Angew. Phys.* **31**, 27.
Nigam, B. P., Sudaresan, M. K. and Wu, T.-Y. (1959) *Phys. Rev.* **115**, 491.
Nodvik, J. S. (1964) *Annals of Physics* **28**, 225.
Nyborg, P. (1962) *Nuovo Cimento* **23**, 1057.
Ohanian, H. C. (1986) *Am. J. Phys.* **54**, 500.
Okun, L. B. (1989) *Physics Today* **42**, 6, 31.
Omnès, R. (1992) *Rev. Mod. Phys.* **64**, 339.
Oms, J., Erman, P. and Hultberg, S. (1969) *Arkiv Fysik* **39**, 573 (1969).
Oppenheimer, J. R. (1964) *The Flying Trapeze: Three Crises for Physicists*, Oxford University Press.
Oppenheimer, J. R. (1970) *Lectures on Electrodynamics*, Gordon and Breach, New York.
Ostriker, J. P. (1971) *Scientific American* **237**, 1, 48.
Pais, A. (n.d.) Rockerfeller University reports *RU82/B/41* and *COO-2232B-91*.
Pais, A. (1972) in *Aspects of Quantum Theory*, A. Salam and E. Wigner (eds), Cambridge Univ. Press.
Pais, A. (1982) *'Subtle is the Lord...'*, Oxford University Press, New York.
Pais, A. (1988) *Inward Bound,* Oxford University Press, New York.
Particle Data Group, *"Review of Particle Properties", Phys. Rev.* **D45**, Part II (1992).
Pasternack, S. (1938) *Phys. Rev.* **54**, 1113.
Paul, W. and Reich, H. (1952) *Z. Physik* **131**, 326.
Pauli, W. (1958) *Theory of Relativity,* Pergamon, Oxford.
Pearle, P. (1982) in *Electromagnetism - Paths to Research,* D. Teplitz (ed), Plenum Press, New York.
Pearle, P., (1977) *Found. Phys.* **7**, 931.
Peitzmann, F. J. and Kessler, J. (1990) *J. Phys. B: At. Mol. Opt. Phys.* **23**, 4005.
Peixoto, E. M. A., Bunge, C. F. and Bonham, R. A. (1969) *Phys. Rev.* **181**, 322.
Penrose, R. (1959) *Proc. Camb. Phil. Soc.* **55**, 137.
Pettus, W.G., Blosser, H. G. and Hereford, F. L. (1956) *Phys. Rev.* **101**, 17.
Pettus, W. G. (1958) *Phys. Rev.* **109**, 1458.
Phipps, T. E. (1962a) *Nature* **195**, 67.
Phipps, T. E. (1962b) *Nature* **196**, 886.
Planck, M. (1907) *Verh. Dtsch. Phys. Ges.* **9**, 301.
Poincaré, H. (1905) *C. R. Acad. Sci.* **140**, 1504.
Poincaré, H. (1906) *Rend. Circ. Mat. Palermo* **21**, 129.
Prokhovnik, S. J. (1967) *The Logic of Special Relativity,* Cambridge University Press.

Pryce, H. M. L. (1938) *Proc. Roy. Soc.* (London) A**168**, 389.
Rafanelli, K. and Schiller, R. (1964) *Phys. Rev.* **135**, B279.
Ranada, A. F. and Vazquez, L. (1984) *J. Phys. A: Math. Gen.* **17**, 2011.
Randels, R. B., Chao, K. T. and Crane, H. R. (1945) *Phys. Rev.* **68**, 64.
Rao, M. V. V. S. and Bharathi, S. M. (1987) *J. Phys. B: Atom. Mol. Phys.* **20**, 1081.
Rasetti, F. and Fermi, E. (1926) *Nuovo Cimento* **3**, 226.
Raskin, P. D. (1978) *Found. Phys.* **8**, 31.
Ray, M. (1965) *Theory of Relativity, Special & General*, Chand, Delhi.
Redhead, M. (1987) *Incompleteness, Nonlocality, and Realism,* Clarendon, Oxford.
Register, D. F., Vuskovic, L. and Trajmar, S. (1986) *J. Phys. B: Atom. Mol. Phys.* **19**, 1685.
Reichenbach, H. (1969) *Axiomatization of the Theory of Relativity*, Univ. of Calif. Press.
Rester, D. H. and Rainwater, J. W., Jr. (1965a) *Phys. Rev.* **138**, A12.
Rester, D. H. and Rainwater, J. W., Jr. (1965b) *Phys. Rev.* **140**, A165.
Rich, A. and Wesley, J. C. (1972) *Rev. Mod. Phys.* **44,** 250.
Rindler, W. (1966) *Special Relativity*, Wiley, New York.
Rindler, W. (1977) *Essential Relativity*, Springer-Verlag, Heidelberg.
Rohrlich, F. (1964) *Phys. Rev. Lett.* **12**, 375.
Rohrlich, F. (1965) *Classical Charged Particles*, Addison-Wesley, Reading.
Rohrlich, F. (1973) in *The Physicist's Conception of Nature*, J. Mehra (ed), Reidel, Dordrecht.
Rohrlich, F. (1982) *Phys. Rev.* **D25**, 3251.
Roman, P. (1985) *Physics Today* **38**, 5, 126.
Rosen, N. (1946) *Phys. Rev.* **70**, 93.
Rosen, N. (1947) *Phys. Rev.* **71**, 54.
Rosenfeld, L. (1973) in *The Physicist's Conception of Nature*, J. Mehra (ed), Reidel, Dordrecht.
Rosser, W. G. V. (1967) *Introductory Relativity,* Butterworths, London.
Rosser, W. G. V. (1968) *Classical Electromagnetism via Relativity,* Plenum, New York.
Ruane, T. F., Waldman, B. and Miller, W. C. (1955) *Phys. Rev.* **98**, 1166a.
Russell, B. (1969) *The ABC of Relativity*, George Allen & Unwin, London.
Ryu, N. (1952a) *J. Phys. Soc. Japan* **7**, 125.
Ryu, N. (1952b) *J. Phys. Soc. Japan* **7**, 130.
Ryu, N. (1953) *J. Phys. Soc. Japan* **8**, 204 (1953).
Saha, B. C., Chaudhuri, J. and Ghosh, A. S. (1976) *J. Phys. Soc. Japan* **41**, 1716.
Salzman, G. and Taub, A. H. (1954) *Phys. Rev.* **95**, 1659.
Schackert, K. (1968) *Z. Physik* **213**, 316.
Schiff, L. I. (1955) *Quantum Mechanics*, Second Edition, McGraw-Hill, New York.
Schmoranzer, H., Grabe, H. and Schiewe, B. (1975) *Appl. Phys. Lett.* **26**, 483.
Schrödinger, E. (1935) *Naturwissenschaften* **23**, 807, 823, 844.
Schroeder, M. (1991) *Fractals, Chaos, Power Laws,* Freeman, New York.
Schulman, L. (1968) *Phys. Rev.* **176**, 1558.
Schulman, L. (1981) *Techniques and Applications of Path Integration,* Wiley, New York.

Schwarzschild, M. (1958) *Structure and Evolution of the Stars*, Princeton Univ. Press.
Schwinger, J. (1948) *Phys. Rev.* **73**, 416.
Scott, M. B. *et al.* (1951) *Phys. Rev.* **84**, 638.
Scott, W. T. (1963) *Rev. Mod. Phys.* **35**, 231.
Shadowitz, A. (1968) *Special Relativity*, Saunders, Philadelphia.
Sherman, N. (1956) *Phys. Rev.* **103**, 1601.
Shull, C. G., Chase, C. T. and Myers, F. E. (1943) *Phys. Rev.* **63**, 29.
Silverman, M. (1982) *Am. J. Phys.* **50**, 251.
Singh, S. and Raghuvanshi, M. S. (1984) *Am. J. Phys.* **52**, 850.
Smith, J. H. (1965) *Introduction to Special Relativity*, Benjamin, New York.
Smythe, W. R. (1939) *Static and Dynamic Electricity*, McGraw-Hill, New York.
Sparnaay, M. J. (1958) *Physica* **24**, 751.
Spiegel, V., Jr., Waldman, B. and Miller, W. C. (1955) *Phys. Rev.* **100**, 1244a.
Spiegel, V., Jr. *et al.* (1959) *Annals of Physics* **6**, 70.
Spivak, P. E., Mikaelyan, L. A., Kutikov, I. Ye. and Apalin, V. F. (1961) *Nucl. Phys.* **23**, 169.
Stephenson, G. and Kilmister, C. W. (1958) *Special Relativity for Physicists*, Longmans, Green, and Co., London.
Synge, J. L. and Griffith, B. A. (1942) *Principles of Mechanics*, McGraw-Hill, New York.
Synge, J. L. (1952) *Nature* **170**, 244.
Synge, J. L. (1965) *Relativity: The Special Theory*, North-Holland, Amsterdam.
Takeno, H. (1952) *Prog. Theor. Phys. (Japan)* **7**, 367.
Tang, F.-C. *et al.* (1988) *Rev. Sci. Instrum.* **59**, 504.
Taylor, J. G. *Special Relativity*, Clarendon, Oxford.
Teitelboim, C., Villarroel, D. and Van Weert, Ch. G. (1980) *Rivista del Nuovo Cimento* **3**, 1.
Terletskii, Y. P. (1968) *Paradoxes in the Theory of Relativity*, Plenum, New York.
Thomas, L. H. (1926) *Nature (London)* **117**, 514.
Tonnelat, M. (1966) *The Principles of Electromagnetic Theory and of Relativity*, Reidel, Dordrecht.
Uhlenbeck, G. E. and Goudsmit, S. A. (1925) *Naturwissenschaften* **13**, 953.
Uhlenbeck, G. E. and Goudsmit, S. A. (1926) *Nature* **117**, 264.
Unruh, W. G. (1976) *Phys. Rev.* **D14**, 870.
Van de Graaff, R. J. *et al.* (1946) *Phys. Rev.* **69**, 452.
Van der Waerden, B. L. (1960) in *Theoretical Physics in the Twentieth Century, A Memorial Volume to Wolfgang Pauli*, M. Fierz and V. F. Weisskopf (eds), Interscience, New York.
Van Duiden, R. J. and Aalders, J. W. G. (1968) *Nucl. Phys.* **A115**, 353.
Van Klinken, J. (1966) *Nucl. Phys.* **75**, 161.
Vuskovic, L., Maleki, L. and Trajmar, S. (1984) *J. Phys. B: Atom. Mol. Phys.* **17**, 2519.
Wellenstein, H. F., Bonham, R. A. and Ulsh, R. C. (1973) *Phys. Rev.* **A8**, 304.
Welton, T. A. (1948) *Phys. Rev.* **74**, 1157.
Wesley, J. C. and Rich, A. (1971) *Phys. Rev.* **A4**, 1341.

Wheeler, J. A. and Feynman, R. P. (1945) *Rev. Mod. Phys.* **17**, 157.
Wheeler, J. A. and Feynman, R. P. (1949) *Rev. Mod. Phys.* **21**, 425.
Wheeler, J. A. (1991) *Scientific American* **257**, 6, 36.
Williams, J. F. and Willis, B. A. (1975a) *J. Phys. B: Atom. Mol. Phys.* **8**, 1670.
Williams, J. F. (1975b) *J. Phys. B: Atom. Mol. Phys.* **8**, 2191.
Williams, J. F. and Crowe, A., Jr. (1975c) *J. Phys. B: Atom. Mol. Phys.* **8**, 2233.
Williams, W., Trajmar, S. and Bozinis, D. (1976) *J. Phys. B: Atom. Mol. Phys.* **9**, 1529.
Zeitler, E. and Olsen, H. (1966) *Z. Naturforschung* **21a**, 1321.
Zurek, W. H. (1991) *Physics Today* **44**, 10, 36.

Index

Fundamental Theories of Physics

Series Editor: Alwyn van der Merwe, *University of Denver, USA*

1. M. Sachs: *General Relativity and Matter.* A Spinor Field Theory from Fermis to Light-Years. With a Foreword by C. Kilmister. 1982 ISBN 90-277-1381-2
2. G.H. Duffey: *A Development of Quantum Mechanics.* Based on Symmetry Considerations. 1985 ISBN 90-277-1587-4
3. S. Diner, D. Fargue, G. Lochak and F. Selleri (eds.): *The Wave-Particle Dualism.* A Tribute to Louis de Broglie on his 90th Birthday. 1984 ISBN 90-277-1664-1
4. E. Prugovečki: *Stochastic Quantum Mechanics and Quantum Spacetime.* A Consistent Unification of Relativity and Quantum Theory based on Stochastic Spaces. 1984; 2nd printing 1986 ISBN 90-277-1617-X
5. D. Hestenes and G. Sobczyk: *Clifford Algebra to Geometric Calculus.* A Unified Language for Mathematics and Physics. 1984 ISBN 90-277-1673-0; Pb (1987) 90-277-2561-6
6. P. Exner: *Open Quantum Systems and Feynman Integrals.* 1985 ISBN 90-277-1678-1
7. L. Mayants: *The Enigma of Probability and Physics.* 1984 ISBN 90-277-1674-9
8. E. Tocaci: *Relativistic Mechanics, Time and Inertia.* Translated from Romanian. Edited and with a Foreword by C.W. Kilmister. 1985 ISBN 90-277-1769-9
9. B. Bertotti, F. de Felice and A. Pascolini (eds.): *General Relativity and Gravitation.* Proceedings of the 10th International Conference (Padova, Italy, 1983). 1984 ISBN 90-277-1819-9
10. G. Tarozzi and A. van der Merwe (eds.): *Open Questions in Quantum Physics.* 1985 ISBN 90-277-1853-9
11. J.V. Narlikar and T. Padmanabhan: *Gravity, Gauge Theories and Quantum Cosmology.* 1986 ISBN 90-277-1948-9
12. G.S. Asanov: *Finsler Geometry, Relativity and Gauge Theories.* 1985 ISBN 90-277-1960-8
13. K. Namsrai: *Nonlocal Quantum Field Theory and Stochastic Quantum Mechanics.* 1986 ISBN 90-277-2001-0
14. C. Ray Smith and W.T. Grandy, Jr. (eds.): *Maximum-Entropy and Bayesian Methods in Inverse Problems.* Proceedings of the 1st and 2nd International Workshop (Laramie, Wyoming, USA). 1985 ISBN 90-277-2074-6
15. D. Hestenes: *New Foundations for Classical Mechanics.* 1986 ISBN 90-277-2090-8; Pb (1987) 90-277-2526-8
16. S.J. Prokhovnik: *Light in Einstein's Universe.* The Role of Energy in Cosmology and Relativity. 1985 ISBN 90-277-2093-2
17. Y.S. Kim and M.E. Noz: *Theory and Applications of the Poincaré Group.* 1986 ISBN 90-277-2141-6
18. M. Sachs: *Quantum Mechanics from General Relativity.* An Approximation for a Theory of Inertia. 1986 ISBN 90-277-2247-1
19. W.T. Grandy, Jr.: *Foundations of Statistical Mechanics.* Vol. I: *Equilibrium Theory.* 1987 ISBN 90-277-2489-X
20. H.-H von Borzeszkowski and H.-J. Treder: *The Meaning of Quantum Gravity.* 1988 ISBN 90-277-2518-7
21. C. Ray Smith and G.J. Erickson (eds.): *Maximum-Entropy and Bayesian Spectral Analysis and Estimation Problems.* Proceedings of the 3rd International Workshop (Laramie, Wyoming, USA, 1983). 1987 ISBN 90-277-2579-9

Fundamental Theories of Physics

22. A.O. Barut and A. van der Merwe (eds.): *Selected Scientific Papers of Alfred Landé. [1888-1975]*. 1988 ISBN 90-277-2594-2
23. W.T. Grandy, Jr.: *Foundations of Statistical Mechanics.* Vol. II: *Nonequilibrium Phenomena.* 1988 ISBN 90-277-2649-3
24. E.I. Bitsakis and C.A. Nicolaides (eds.): *The Concept of Probability.* Proceedings of the Delphi Conference (Delphi, Greece, 1987). 1989 ISBN 90-277-2679-5
25. A. van der Merwe, F. Selleri and G. Tarozzi (eds.): *Microphysical Reality and Quantum Formalism, Vol. 1.* Proceedings of the International Conference (Urbino, Italy, 1985). 1988 ISBN 90-277-2683-3
26. A. van der Merwe, F. Selleri and G. Tarozzi (eds.): *Microphysical Reality and Quantum Formalism, Vol. 2.* Proceedings of the International Conference (Urbino, Italy, 1985). 1988 ISBN 90-277-2684-1
27. I.D. Novikov and V.P. Frolov: *Physics of Black Holes.* 1989 ISBN 90-277-2685-X
28. G. Tarozzi and A. van der Merwe (eds.): *The Nature of Quantum Paradoxes.* Italian Studies in the Foundations and Philosophy of Modern Physics. 1988 ISBN 90-277-2703-1
29. B.R. Iyer, N. Mukunda and C.V. Vishveshwara (eds.): *Gravitation, Gauge Theories and the Early Universe.* 1989 ISBN 90-277-2710-4
30. H. Mark and L. Wood (eds.): *Energy in Physics, War and Peace.* A Festschrift celebrating Edward Teller's 80th Birthday. 1988 ISBN 90-277-2775-9
31. G.J. Erickson and C.R. Smith (eds.): *Maximum-Entropy and Bayesian Methods in Science and Engineering.* Vol. I: *Foundations.* 1988 ISBN 90-277-2793-7
32. G.J. Erickson and C.R. Smith (eds.): *Maximum-Entropy and Bayesian Methods in Science and Engineering.* Vol. II: *Applications.* 1988 ISBN 90-277-2794-5
33. M.E. Noz and Y.S. Kim (eds.): *Special Relativity and Quantum Theory.* A Collection of Papers on the Poincaré Group. 1988 ISBN 90-277-2799-6
34. I.Yu. Kobzarev and Yu.I. Manin: *Elementary Particles. Mathematics, Physics and Philosophy.* 1989 ISBN 0-7923-0098-X
35. F. Selleri: *Quantum Paradoxes and Physical Reality.* 1990 ISBN 0-7923-0253-2
36. J. Skilling (ed.): *Maximum-Entropy and Bayesian Methods.* Proceedings of the 8th International Workshop (Cambridge, UK, 1988). 1989 ISBN 0-7923-0224-9
37. M. Kafatos (ed.): *Bell's Theorem, Quantum Theory and Conceptions of the Universe.* 1989 ISBN 0-7923-0496-9
38. Yu.A. Izyumov and V.N. Syromyatnikov: *Phase Transitions and Crystal Symmetry.* 1990 ISBN 0-7923-0542-6
39. P.F. Fougère (ed.): *Maximum-Entropy and Bayesian Methods.* Proceedings of the 9th International Workshop (Dartmouth, Massachusetts, USA, 1989). 1990 ISBN 0-7923-0928-6
40. L. de Broglie: *Heisenberg's Uncertainties and the Probabilistic Interpretation of Wave Mechanics.* With Critical Notes of the Author. 1990 ISBN 0-7923-0929-4
41. W.T. Grandy, Jr.: *Relativistic Quantum Mechanics of Leptons and Fields.* 1991 ISBN 0-7923-1049-7
42. Yu.L. Klimontovich: *Turbulent Motion and the Structure of Chaos.* A New Approach to the Statistical Theory of Open Systems. 1991 ISBN 0-7923-1114-0

43. W.T. Grandy, Jr. and L.H. Schick (eds.): *Maximum-Entropy and Bayesian Methods.* Proceedings of the 10th International Workshop (Laramie, Wyoming, USA, 1990). 1991 ISBN 0-7923-1140-X
44. P.Pták and S. Pulmannová: *Orthomodular Structures as Quantum Logics.* Intrinsic Properties, State Space and Probabilistic Topics. 1991 ISBN 0-7923-1207-4
45. D. Hestenes and A. Weingartshofer (eds.): *The Electron.* New Theory and Experiment. 1991 ISBN 0-7923-1356-9
46. P.P.J.M. Schram: *Kinetic Theory of Gases and Plasmas.* 1991 ISBN 0-7923-1392-5
47. A. Micali, R. Boudet and J. Helmstetter (eds.): *Clifford Algebras and their Applications in Mathematical Physics.* 1992 ISBN 0-7923-1623-1
48. E. Prugovečki: *Quantum Geometry.* A Framework for Quantum General Relativity. 1992 ISBN 0-7923-1640-1
49. M. H. Mac Gregor: *The Enigmatic Electron.* 1992 ISBN 0-7923-1982-6

KLUWER ACADEMIC PUBLISHERS – DORDRECHT / BOSTON / LONDON

Printed in the United States
27863LVS00002B/277